Mario Quintas Neto

Environmental education in the municipality of Palmas/TO

Mario Quintas Neto

Environmental education in the municipality of Palmas/TO

Between practices and representations

ScienciaScripts

Imprint

Cover image: www.ingimage.com

This book is a translation from the original published under ISBN 978-613-9-73127-5.

Publisher:
Sciencia Scripts
is a trademark of
Dodo Books Indian Ocean Ltd. and OmniScriptum S.R.L publishing group

120 High Road, East Finchley, London, N2 9ED, United Kingdom
Str. Armeneasca 28/1, office 1, Chisinau MD-2012, Republic of Moldova, Europe
Printed at: see last page
ISBN: 978-620-7-89573-1

ACKNOWLEDGEMENTS

The act of saying thank you, here dressed up in formality, does not do justice to all the opportunities, support and complicity that I have had the privilege of being honoured with throughout the course of this study. But there are certainly people who need to be recognised and thanked, because without them, this moment would not be possible.

Firstly, I'd like to thank my family and, for many of you like me who are expats, you understand how broad and all-encompassing this concept can be. So I'd like to thank my family from Tocantins as much as my blood family. To the latter for their support and unconditional love, even though I am a prodigal son and have been away from home for so long, and to the former for having allowed me to create ties and bonds with a land so distant and different from my own.

I would like to say a huge thank you to my advisor, Prof Dr Alberto Akama, for being an outstanding professional, assertive, objective and precise, helping in a coherent and efficient way, regardless of geographical distance. Thank you very much for your generosity in guiding me and showing me the paths to follow and the pitfalls to avoid. I would also like to thank the professors who made up the panel, who in one way or another contributed to the construction and improvement of this study.

I would like to thank the Federal University of Tocantins, and in particular the teachers of the Postgraduate Programme in Environmental Sciences for always welcoming me so well into the course, helping me, awakening me and teaching me how to adjust the lenses of interdisciplinarity. My special thanks go to Professor Dr Bertolin, a true friend on this journey.

To the many colleagues from the master's and doctoral programmes for sharing visions and views, socialising knowledge, expectations and anxieties.

Thanks to P.A., for never again being sure whether it's better to study or potatoes. To Bibiana, for agreeing to embark on this journey with me. To Aleca, for questioning academia. To Rodrigo and the rest of the people in the data or on the boards for being there when I needed them and when I didn't either. To my parents and brothers, for understanding that bad Christmas. Thank you all.

SUMMARY

Environmental education today represents a path towards balance and a relationship between society and the environment. It was in this context, taking up the nuances of the environmentalist discourses of the 1960s and 1970s, that EE was given a special role in the search for a sustainable society. In this sense, when the capital of Tocantins, Palmas, was planned and built in the middle of the Cerrado, the whole discourse was geared towards the construction of an ecological and environmentally friendly capital. Based on this premise, this study sought to identify the social and environmental representations at work in the context of the formation and development of Palmas, as well as how these representations were able to transform practices related to the environment and environmental awareness, especially through environmental education. In order to identify the discourses, the city's formation history and projects were analysed, as well as municipal legislation related to the Master Plan, the environment and environmental education. The data obtained through the bibliographical review and the sources consulted was processed through discourse analysis, which used environmental legal hermeneutics to understand how Environmental Education was treated within the discourses of the representations and how these discourses, which gave an account of Palmas - the Ecological Capital - were effectively put into practice. During the investigation, the use of ecology and an environmentally responsible and sustainable discourse by the representations became evident, especially by the public authorities, while at the same time very few of the concerns raised were addressed. There was also evidence of a disregard for the effectiveness and sustainability of environmental education projects, which were carried out in a watertight, disconnected manner, to meet specific and momentary ends. Three important time periods were divided to analyse the discourses and representations, each covering around a decade of Palmas' existence and with significant improvements in the legislation that accompanied urban development. The role of the urban space in the gap between the practices and discourse of the existing representations was also highlighted, as was the emergence of a new approach, a new programme called Sustainable Palmas, which is still in progress and which takes up the discourse of Palmas as the Ecological Capital of the city's beginnings, although there has been no change in the intentionality and guarantee of efficiency and continuity of the Environmental Education projects.

Keywords: Environmental Education. Environmental Representations. Discourse Analysis. Environment.

SUMMARY

CHAPTER 1

INTRODUCTION

Environmental representations have permeated the process of establishing and developing the municipality of Palmas and the state of Tocantins since its beginnings, playing a prominent role in political discourses and practices. The adoption of the title of Ecological Capital of Brazil and the 1988 Federal Constitution had a significant influence in legitimising the discourses and representations of power that justified the way in which the process of creating and establishing the new federal unit and its capital took place.

The municipality was founded in 1989, a year after the emancipation of the state of Tocantins, and was created with the aim of consolidating the region's political and economic independence from the south of Pará and the north of Goiás. Influenced by the architectural and political plans for Brasilia, and under the umbrella of environmental and democratic movements, the creation of the new capital required a discourse in keeping with the country's historical and social situation.

The dominant group in Tocantins did not want to maintain the existing geographical and political configuration. What was needed was something new, in order to establish a difference and give strength and substance to the developmentalist and progressive discourse that enabled the territory's independence. So,

> the option to build a new city was gradually asserted. The strategy, as well as seeking to escape the political disputes consolidated in the cities chosen as alternatives during the selection process, aimed to intervene in a positive way in the organisation of the territorial space of Tocantins (TEIXEIRA, 2009, p.92).

But simply building an artificial metropolis in the middle of the savannah, like Brasilia, was not economically or morally justifiable. It was necessary to adopt a discourse of intentionality that justified the work not only from the point of view of political and economic independence, but also from the point of view of sustainability, forcing a spirit that was in keeping with its time.

It was in this sense that political and economic discourses were moulded with a view to respecting and integrating the environment and taking sustainability as a central point. This shaped a city that needed, as it grew, to work through the dichotomies between development discourse and practice and environmental preservation discourse and practice.

The environment is assumed to be a holistic and ecological concept, a conception of the world as an integrated unit, a collection of various elements brought together, sometimes with a fundamental interdependent relationship between all the parties involved (CAPRA, 2002).

Based on this assumption, the different discourses pay attention to the different interests of the representations active in the constitution and development of Palmas. The different players who have acted and are acting in the development of the capital obviously often have contrasting interests, but they are not inseparable.

The decisions that led to the formation of Palmas had to be justified in terms of the configuration and the **way in which it presented itself as an "ecological" capital. In this sense,** the quest for sustainability had to find support in the discourses and models of capitalist production, as it was necessary to justify advances and social behaviour based on preservation, conservation and the guarantee of future rights. At the same time, it sought to guarantee environmental commodification and unify discourse with existing practice (CORNETTA, 2012).

Capital played an important role in the formation of Palmas, following the process described by Correa (1996), especially with the presence and action of social agents and their relations with the urban space that also influence these dynamics: owners of the means of production, landowners, property developers, the state and excluded social groups, all acting in their own interests in a constant dispute to occupy the urban fabric. This has led to conflicts, as the sectors with a monetary interest have acted to prevent the occupation of this network by those who would not bring the desired profits from the use of the land.

In this way, Palmas was born as a city that was planned in its discourse and project, but executed in a different way, catering to different interests, especially those of the representations that were at work in its formation. It is these groups and representations that, while moulding the city, have their discourses of justification influenced by the socio-environmental movements of previous decades. The definitions and recommendations of the Club of Rome, the Stockholm Conference of 1972, the New Charter of Athens and the Brazilian Federal Constitution of 1988 and pre-Rio 92 were directly reflected in the planning and representations constructed and adopted by the groups articulated around the new territory.

Thus, at the time of its creation, there was a socio-environmental discourse committed to sustainability and the preservation of the environment. This discourse can be understood as a locus for the relationship between individuals and society and nature, in an integrated, inclusive way, with various socio-legal aspects, but essentially a social space for the relationship between nature and society.

According to Chartier (1990), social representations were fundamental to understanding the unity and conformation of the discourses of power groups and the intentions behind these discourses. The author also explains that political and legal discourses have in their essence a set of values that act to the detriment of others in a dispute over symbolic power, which complements the issue of capital in disputes defended by Bourdieu (1998), especially in relation to identity and power. Lefbvre

(1983) also adds that these representations influence existing social practices while interpreting and perpetuating them.

For environmental representations, it was necessary to understand the intertwined ethical and political values, which sought to transform social paradigms and their relationship with nature, as well as the maintenance of this relationship, together with the interdependence of the actors (REIGOTA, 1994).

As a result of the discourse adopted at its creation, the municipality had to commit itself to the environment. Thus, on the recommendation of the Stockholm Conference, significant importance was attached to Environmental Education, both at municipal and state level, which emerged as an element for understanding and interpreting existing representations, and how practices do or do not go beyond the discourse emanating from them.

According to Fiorillo (2008), environmental education means reducing environmental costs, implementing the principle of prevention, developing ecological awareness and preserving indivisible and diffuse resources, i.e. those belonging to the general public. This understanding complements the one set out and adopted by the Ministry of the Environment, which defines environmental education as a permanent process in which individuals and the community become aware of their environment and acquire knowledge, values and skills with the aim of preserving and protecting the environment (ADAMS, 2005).

This study was born out of this dispute between discourse and practice. It sought to understand, in an interdisciplinary way, the historical process of the capital's formation, as well as the development of legislation and environmentalist-educational discourse and environmental education practices in the municipal sphere over the 25 years of its existence.

The geographical and historical focus was determined by the unique characteristics of the place: being a young capital, formed from a supposed geographical void, with artificial growth, in theory planned, and being the capital of one of the last states created in Brazil after the 1988 Constitution.

The time frame, at first glance extensive, covered the embryonic phase, even before the city was founded, right up to the present day. This was necessary in order to understand the historical process leading to the current legal-pedagogical context. The period after the founding of the city could be divided into at least three distinct phases: the initial years, which correspond to the decade from 1990 to 2001, approximately; a second phase that comprised the 2000s, when legislative changes regarding Environmental Education, the Convention on Biological Diversity (CBD), the Law of Guidelines and Bases (LDB) and other educational documents and, finally, the last ten years, when a developmentalist environmental discourse was resumed.

It is no coincidence that these three phases coincided with the original planning and the Municipal Organic Law, with the revision of the Master Plan in 2005 and, recently, with the new revision of the Master Plan and the launch of the Sustainable Palmas programme.

The intention of the study was not to verify the representation or perception of educators or schools in Palmas within the context of Environmental Education, which has been exhaustively worked on by other researchers. This study focused on the political-pedagogical bias that was based on a discourse of environmentalism and environmental education/awareness to legitimise its practices, even when these did not coincide with the ideals defended. Therefore, although the pedagogical plans were analysed, it was more interesting to look at the legal provisions and the projects that were actually proposed or put into practice.

We then sought to analyse the construction of the developmentalist political discourse in the light of environmental education in the municipality of Palmas. This analysis was based on legislative documents and educational policy practices, from the occupation of the urban space to the current and future prospects for the region,

To this end, a documentary bibliographical survey was carried out to choose the guiding concepts, as well as to define the methodology. This culminated in the choice of the bibliographical review and the hermeneutic analysis of the environmental discourse present both in the legislation and in the other documents of the acts carried out.

It was therefore necessary to analyse Palmas's environmental and educational urban legislation in order to understand current practices and understandings. Likewise, the Municipal Education Plan (PME) and the projects developed over the years by the various municipal bodies responsible for the environment and environmental education in Palmas were analysed. We also analysed the minutes and documents that made up the processes of revising the master plans, including the meetings with the community, allowing us to see the perception of these groups, which also sometimes contradicts the official discourse.

In the first part of the study, the theoretical framework adopted was discussed, defining the understanding of the environment as a social construct, based on authors such as Springer (2010), Harré (1999), Drummond (1991), Gonçalves (2006), Santos (1996), Ellict (2002), Grinover (1984), Farhi (2006), Cornetta (2012), among others. Environmental education was also defined from a pedagogical, legal and environmental perspective, using authors such as Moura (2002), Lago (2007), Capra (2000), Clapp (2005), Aires (2010) and others.

Social and environmental representations and other relevant terms such as field, habitus and symbolic capital were conceptualised, based on authors such as Chartier (1990), Bourdieu (1998), Moscovici (1978), Sachs (2002), Bruger (1996) and Reigota (1994). The understanding of urban space, its configuration and occupation was subsidised by Correa (1989), Mumford (1998), Santos (1985), Lefebvre (1999) and others. Finally, the concepts formulated by Lyra (2000), Rangel (2012), Padilha (2010) and Fiorillo (2008) were used to address the environment and the law.

Secondly, the methodology, definitions of the object and research techniques used in the study were presented. Emphasis was placed on hermeneutic discourse analysis, based on Foucault (1996), Robles (2005), Belchior (2009), Reigota (2007), Leff (2003) and other authors.

The following sections analyse the objects of study, both the state of Tocantins and the municipality of Palmas, in terms of their historical and geographical aspects, and from a legislative perspective. The projects, urban planning proposals, fundamental laws and speeches issued during that period were analysed. In addition, the proposals involving the Master Plan and the revision of this document were analysed from different possible perspectives, with special attention to the transformations due to the Cities Statute.

The last stage of the study discusses environmental educational representations and practices in Palmas. In this way, the constitution of the urban space presented in the previous chapter is related to the legislation and transformations that have taken place in the city throughout its history. The intention in this case is to contribute and verify which environmental education practices have gone beyond the official discourse. Finally, the proposal of the Sustainable Palmas programme and the analysis of the resumption of a discourse committed to development and environmental sustainability are presented and discussed.

As a critique, the problem faced in this dissertation is finally defined: **"How does Environmental Education deal with the result of disputes between political discourse and** environmental practices developed in the municipality of Palmas in Tocantins from its formation **to the present day?".**

CHAPTER 2

THEORETICAL BACKGROUND

2.1 The environment as a social construct

The understanding of the environment permeates each individual's perception in a different way, varying greatly according to the background and experience of the interlocutor. It is therefore necessary to understand and define the concept of the environment used in this work.

Springer (2010) questions the effectiveness of environmental studies in the absence of clarity about how nature and the environment are understood. In this sense:

> environmental problems are fundamentally problems of knowledge. This has strong implications for any environmental policy - which must include a knowledge policy - and also for education. Understanding environmental complexity is not a problem of learning about the environment, but of understanding knowledge about the environment (LEFF **apud** SPRINGER, 2010, p. 169).

The conceptualisation of Nature, the Environment and the Environment varies according to the research perspective, and is different for the areas of science that deal with these topics. For a biologist, nature is the study of biotic and abiotic components, but it overlooks the cultural and social interactions that emanate from it. On the other hand, the humanities relegate the study of biotic and abiotic components to the background, while economists have a market-based view of nature.

In order for the study to fulfil its objectives, it is important to understand the concepts necessary for the development of the work. This will allow for a holistic approach that permeates all areas of science related to the topic. Knowing or defining what was meant by the environment at the time of the founding and planning of the municipality of Palmas will allow us to move beyond the environmentalist discourse, which is deeply rooted in the naturalistic view. In this way, it will be possible to verify the praxis and effectiveness of the proposals launched on the subject of Environmental Education.

2.2 Nature: Definitions and Concepts

Harré (1999) points out that the speeches made will vary according to the intentionality and perception of each of the social actors involved, and may involve issues on different scales, ranging from micro-scale issues to global problems such as global warming and the reduction of the ozone layer.

In addition to the question of scale, there are a multitude of different themes, from economic discourses, to rationalism, sustainable development, the integration of man and nature, among others

(MOL, 2003).

The attempt to understand the environment not just as nature means that we have to look to geography and other sciences to understand the whole. However, we believe that defining nature seems to be an adequate starting point, since one concept refers to the other in terms of Western thought and ethics.

Drummond (1991), when working on environmental history, finds a similar dilemma when it comes to defining what is understood by nature and what is understood by the environment:

> First comes Leo Marx, actually a historian of literature. In the early 1960s, he produced an extremely influential book among environmental historians: **The machine in the garden-technology and the pastoral ideal in America** (London, Oxford University Press, 1964). Marx taught the entire "first generation" of environmental historians the analytical importance of the values attributed to different forms of nature (DRUMMOND, 1991, p. 13).

He also points out that in the West there is a dichotomy between "civilised" and "wild" nature, and that it is inherent in Western culture to consider man-made and maintained gardens and nature to be beautiful, as opposed to the chaotic and wild, untouched nature that arouses society's civilising zeal (DRUMMOND, 1991).

As a cultural construction, the meaning of nature appears in the work of Porto Gonçalves, where he scrutinises, based on Milton Santos, the concept of nature to be adopted by Geography:

> In our society, nature is defined by what opposes culture. Culture is seen as something superior that has managed to control and dominate nature. That's why the Neolithic revolution, or agriCULTURE, is seen as a milestone in history, since with it man went from gathering what nature had to hand to cultivating it.
>
> **"Naturally" is used to collect what is** planted and cultivated (GONÇALVES, 2006 p.26).

He also explains that nature, in the eyes of our society, is an object to be dominated by man. However, at the same time, he recognises that not all men can dispose of or appropriate it, a critical reference to the current economic system. Thus, he points out the great paradox of modern man in his attempt to establish an anthropocentric relationship - an action that leads to the submission of the other to whom he relates (GONÇALVES, 2006).

He then traces the construction of the concept of nature back to the pre-Socratics and Greco-Roman currents, passing through Judeo-Christian thought to modernity, which, through modernist and Cartesian thought, created pragmatism and consolidated the man-nature dichotomy. Crowning the construction of the term and the relationship between man and nature, Gonçalves states:

> The idea of nature as objective and external to man, which presupposes an idea of man as unnatural and outside of nature, crystallised with the industrial civilisation ushered in by capitalism. The sciences of nature are separated from the sciences of man; a colossal gulf is created between one and the other and, as we will see later, all of this is just a conception of the world. Ecology as knowledge and, above all, the ecological movement are trying to denounce the consequences of these perceptions, although they often do so permeated by the principles and values of their detractors [...] (GONÇALVES, 2006, p.35).

Likewise, Springer points out the construction of the concept of nature for Geography, following a similar path to Gonçalves, listing four important phases of conceptualisation:

> In order to characterise the various ways of conceiving Nature over time (Western society), historical facts, philosophical knowledge, scientific knowledge and the conception of Nature were interwoven. Some milestones (historical, theoretical and epistemological) defining periods were identified, and their characterisation demonstrates throughout the text the intrinsic and inseparable relationship between history, philosophy, scientific knowledge and the conception of Nature. Four typologies were created: a) Phisis in Ancient Greece; b) divinised Nature; c) mechanical and rational Nature and d) new theories: what is Nature? (SPRINGER, 2010, p. 161).

In order to explain the concept of nature from the perspective of the relationship between man/society and nature, Springer concludes that "Nature therefore becomes a space, of which we will have a conceptual extension of the familiar and everyday spaces given by man's direct experience [...]" (TUAN *apud* SPRINGER, 2010, p. 166).

In the wake of the materialist current, Gonçalves (2006) presents nature in its pragmatic and utilitarian conception as a place and resource at the disposal of modern, industrial man. Referring to Milton Santos' concept, we see the transformation of the bucolic and despicable nature found in Greek and Judeo-Christian philosophical thought into a nature that serves the interests of labour and capital, giving it a value that was previously despised.

> **When the** "Environment", **as** "Spectacle Nature", replaces Historical Nature, the place where all men work, and when **"cybernetic" or "synthetic" nature replaces the analytical nature of the past,** the process of concealing the meaning of history reaches its peak. It is also in this way that a painful confusion is established between technical systems, nature, society, culture and morality (SANTOS, 1992, p. 102).

Peluso (2013) gives clues as to how Milton Santos defined the concept of nature, which is understood as the set of characteristics and elements that make it up. For him, "the essence of space is social. In this case, space cannot just be made up of things, geographical objects, natural and artificial, the whole of which gives us Nature" **(SANTOS** *apudPELUSO,* 2013, p. 23).

In agreement with Gonçalves (2006), the author states that the concept of nature in the work of Milton Santos has undergone a transformation, recognising the transformation from the magical place of early philosophy to the pragmatism of modernity, seeing nature as a source of resources for work:

> The concept of nature has changed; it has ceased to be a subject - the nature of primordial times that we will never know, and has acquired the condition of object - the nature of the current urban period, in which the city is the most perfect place for the predominance of techniques. In other words, it has ceased to be First Nature and has become second nature forever. Santos (1997, p. 203) writes that the old distinction between first nature and second **nature must be made more flexible, since "nature already modified by** man is **also first nature", since "production is no longer the action of** labour on nature, but labour on labour (PELUSO, 2013, p.25).

As we have seen, the concept of nature within geography has been moulded as it has evolved, allowing for at least three philosophical shades of understanding. The first, positivist and neopositivist, understands nature as an entity external to man, whose function is to provide the necessary resources for man to live and develop. This current is rationalist in nature and is concerned with the preservation, planning, rational use and management of natural resources (SPRINGER, 2010).

The second current seeks meaning in an attempt to establish and understand the relationship between individuals and society and the natural environment and its representation. In this way, it seeks to identify the elements that make it up and interpret how they relate to each other and to the social groups that interact with it.

The third current, of dialectical materialist origin with a critical perspective, understands nature as a resource that serves as a backdrop for other aspects, especially economic and political aspects that point to the transformation and limitation of nature as a resource that is no longer accessible to all groups and individuals, but a precious commodity that is only available to those **who can "afford" the "green" (SPRINGER, 2010).**

Nature is a global economic resource that influences conflicting political issues. While on the one hand there are concerns about democratic freedoms, on the other there are neoliberal economic interests, which in many ways diverge, especially when they relativise rights and the need to conserve and preserve natural resources. Nature therefore becomes a resource, a source of social influence (ELLIOTT, 2002).

For this study, it seems more appropriate to use the concept according to the second current presented here, which establishes and seeks to understand the relationship between individuals and society and the natural environment and its representation. This, combined with Cassirer's understanding **that "the** symbolic dimension makes us live in a new dimension of reality, since man lives in a symbolic environment" (CASSIRER, 1994, p. 48).

2.3 The Environment: Complexity and Implications

Once the discussion about nature is over, we look at the environment. To Santos (1996), a geographical space:

> By understanding this geographical content of everyday life, we can perhaps contribute to the necessary understanding (and perhaps theorisation) of this relationship between space and social movements, seeing in materiality this indispensable component of geographical space, which is, at the same time, a condition for action; a control structure, a limit to action; an invitation to action. We do nothing today that is not based on the objects that surround us (SANTOS, 1996, p.257).

For law as a pragmatic and technical science, there is a doctrinal acceptance of the concept of the term, according to which

> the word environment indicates the place, site, enclosure, space that surrounds living beings or things. The expression Environment, although redundant (because the word **environment** already includes the notion of surroundings), ended up being enshrined among us (GONÇALVES, 2003, p.86).

The environment is also treated as a unitary and indivisible legal asset protected by the Brazilian Constitution, defined as the sum of the various legal assets that make it up. It has a neopositivist definition that is characteristic of the Brazilian legal system, and is defined as

> an autonomous and unitary legal asset, which cannot be confused with the various legal assets that make it up. It is not a simple sum of flora and fauna, water resources and mineral resources. It results from the suppression of all the components that can be identified in isolation, such as forests, animals, air, etc. The environment is therefore a **res communes omnium**, a thing common to all, which can be made up of assets belonging to the public or private domain (ANTUNES, 2004, p.240-241).

This emphasises the universalist, but at the same time multiple and comprehensive nature of the protected good conferred by the Federal Constitution. For law as a science, the environment is defined as a set of goods related to both space and things that require protection.

However, it is not limited to being the collective compound of these objects and other goods, as it must also be inherent to all the individuals who can benefit from it.

It is, therefore, a diffuse right, a key element in understanding the environment for the Law, since it requires the good to go beyond the right of a single individual, to be indivisible as a right and to satisfy the indeterminate community. However, it relates to the circumstance itself, and Grinover (1984) defines diffuse rights in this way and gives an example:

> [...] it comprises interests that do not find support in a well-defined basic relationship, reducing the link between people to circumstantial or extremely generic factors, to factual data that are often accidental or changeable: living in the same region, consuming the same product, living under certain socio-economic conditions, being subject to certain undertakings, etc. (GRINOVER, 1984, p.30).

Thus, the definition of the environment in terms of the law is portrayed as a set of legally protected goods that do not find support in a basic good relationship.

defined. But at the same time, it impacts and refers to individuals as a whole, constituting per se a diffuse right that encompasses a series of other individual goods under its indivisible protection.

Rescuing the concept of the milieu is important for understanding what will be used as a field, for the use of representations, so that they don't live in places apart from the society in which they manifest verisimilitude. For Bourdieu, the milieu is the space in which discourse and social action are disputed, discourse being an instrument of power with its own symbolic meaning and force. Therefore, the concept of field can be understood as a locus of social action, where power relations are manifested, where social actors dispute over specific interests. In other words, the place where society meets to dispute interests (BOURDIEU, 2004).

The concept of the environment, rescued from the ecological movements, was established on the basis of the changes and social movements that permeated humanity between the 1940s and 1960s, in the post-war period. A cultural, ideological and practical revolution or counter-revolution was breaking out in humanity, laying foundations and establishing new and important movements, above all for rights and equality.

It is important to establish that in the post-war period, which brought paradigm shifts in its wake, there was a distancing and at the same time a rejection of the attempt to explain the Social Sciences and Culture through the Natural Sciences:

> **Perhaps the rejection of the colonialist and/or warmongering implications of Herbert Spencer's "social Darwinism" and the geographical and racial determinisms of the** 19th century was the most important cause of this. The nascent social sciences, in all their now established guises, fought against these reductionisms and rejected explanations of culture through biology or **any natural science. They resisted the concepts and models of "natural history" and its heirs, including *the* science of ecology.** Human societies, especially modern ones, were conceptualised and investigated under the assumption that they only obeyed the hands of cultural time over a few millennia (DRUMMOND, 1991, p.3).

For much of the 20th century, social scientists were reluctant to admit that human society was linked to nature. This stoic disconnection remained unshakeable both because of the fragility of the multiple theoretical corpus of the human sciences and for historical reasons. Due to the paradigms and problems arising from the fascist and extremist ideologies that plagued Europe and the world, there was a preoccupation, conscious or not, that ended up isolating culture from nature:

> Two American sociologists, W Catton and R. Dunlap (1980), asking in the 1980s why the social sciences had not been at the **forefront of the global "ecological awakening" of the previous decade, concluded** that, since their inception, they had adopted **a "*human exemptionalism* paradigm"** to the factors of nature. In this paradigm, each society and human culture in general are intelligible only in themselves. In Durkheim's terms, social facts can only be explained by other social facts. This is why, according to Catton and Dunlap, for more than a century sociologists - and other social scientists, they add - have carefully avoided suggesting that culture is in any way limited or conditioned by natural factors (DRUMMOND, 1991, p.3-4).

The environmentalist social movements of the late 20th century, together with certain current environmental crises, called for the return of the social sciences to a look beyond humanism. This removed them from the isolation in which they had placed human society, detached from nature (DRUMMOND, 1991).

It was necessary to consider the environment as a holistic element that therefore encompassed various geographical, historical, social and natural aspects:

> All the elements that make up the place, the space, where human beings live form the Environment: the soil, the housing, the climate, the roads, the avenues etc. (VESENTINI, 2008, p. 326).

Thus, we don't talk about the environment as a geographical space, but rather the space where representations vie for consolidation through discourse and practice. The complexity demanded of the different sciences has meant that both the singular and the universal have had to be addressed within the natural and social sciences. In this sense, Morin, when dealing with ecology and its complexity within the Biological Sciences, illustrates the need to go beyond the isolation of the singular towards the universal in a concatenated way:

> Biology today does not conceive of the species as a general framework of which the individual is a singular case. It sees the living species as a singularity that produces singularities [...] The development of the ecological discipline in the biological sciences shows that it is within the localised framework of ecosystems that singular individuals develop and live. Therefore, we must not replace the singular and the local with the universal: on the contrary, we must unite them (MORIN, 2005, p.178-179).

Morin (2005) says that complexity cannot be understood as a static recipe, but rather as an incentive to think beyond the limitations imposed by the boundaries of the sciences, because it is the incompleteness of knowledge. He points out that the human characteristic, its complexity, its multifaceted formation, with biological, cultural, psychological, historical, cultural, physical and spiritual aspects, must be understood and analysed on the basis of complex thinking, through which it will be possible to establish identity and differences that would otherwise end up sublimating several of these aspects due to a reductionist process. Complex thinking has made it possible to see man and his relationships with the environment as multifaceted objects that are necessary for understanding his different perceptions and aspects:

> For a long time, and perhaps even today, it was believed that the mistake of the human and social sciences was that they couldn't get rid of the apparent complexity of human phenomena in order to elevate themselves to the dignity of the natural sciences that made simple laws, simple principles and managed to ensure that the order of determinism reigned in their conceptions (MORIN, 2005, p.177).

The relationship between Man, Culture, Nature and the Environment, however, goes beyond the biological determinism explained above. Laraia (2001) points out that the cultural bias, which influences and complements the geographical, political and social bias, also plays a part in the process of constructing conceptions.

Therefore, the environment is understood as a holistic and ecological concept, according to which the world is understood as an integrated unit, a collection of diverse elements brought together.

It is important to understand the environment through the philosophical current that understands it as "ecological", distinguishing between two ecologies, one shallow and the other deep. Shallow ecology is centred on man, inserted in or separated from nature, determining the latter's mere instrumental value. Deep ecology cannot disassociate nature and man. On the contrary, it considers that all living beings are interdependent, connected through the web of life, as in Capra's studies (SUITER, 2009).

Farhi Neto (2006), in his study of environmental philosophical conceptions, points to the existence of distinct philosophical currents, which he groups under three headings: liberal environmentalism, libertarian environmentalism and social ecology. These three headings concatenate the different currents, grouped according to the proximity of their thinking. In this study, the approach to the liberal environmentalist current seems more appropriate. At this point, Farhi indicates the existence of three currents that belong to liberal environmentalism, and the theses defended by the political liberals represented by de-Shalit are of interest.

In this current, the main concern is the preservation of the liberal tradition in order to avoid totalitarianism that would oppose the principles of the current: freedom, equality and justice in the distribution of goods. The activists of political liberal environmentalism argue that the necessary environmental measures can come from a political consensus through debate, without the need to curb individual freedoms (FARHI, 2006).

The political liberalism proposed by de-Shalit cannot be confused with economic liberalism, which has principles that run counter to environmental preservation, since it is not economically viable (FARHI, 2006). This is evidenced by the pursuit of profitability and the peculiar dynamics of the current system of production and consumption. Making the environment a product was a logical consequence for the capitalist system of the Second Industrial Revolution (CORNETTA, 2012).

The quest for sustainability became the keynote of the discourse of the capitalist production model at the end of the 20th century, and began to dictate social progress and behaviour. Not only is it a product whose political discourse is based on preservation, conservation and guaranteeing future rights. There is also a technical priority to prove the need for this environmental commodification. Economic liberalism therefore needs to transform its object and objective in order to remain congruent with its principles (CORNETTA, 2012).

Returning to de-Shalit, the environmental problem needs to be dealt with urgently, as it is incapable of waiting for the self-corrective action of the market. According to de-Shalit, there is no evidence that environmental issues can be solved by economic means. He also states that simply considering the individual as Homo economicus is not enough to promote the resolution of environmental problems; it is necessary to see man as a political subject (DE-SHALIT apud FAHRI, 2006).

Therefore, de-Shalit lists four fundamental aspects of political liberalism that genuinely characterise the relationship between liberalism and environmental issues: 1) the principle of equality; 2) a commitment to tolerance of diversity and the coexistence of opposing theses; 3) the defence of individual rights; 4) an internationalist vision (FAHRI, 2006).

In the first respect, equality between individuals determines respect for one's neighbour, recognition of the other as equal and not as different or submissive. This principle favours environmentalism by recognising in other living beings an equal right to the environment and the preservation of ecosystems.

The second principle, which refers to coexistence with diverse theses, manifests itself in tolerance of disagreement and divergent ideas, accepting the other person's view as a possible solution, broadening dialogue and allowing diverse conceptions of ideas to flourish (FAHRI, 2006).

The third political liberal assumption relates to the overriding right of the individual over institutions, in an appreciation of the principle of justice and the preservation of individual freedoms and rights. This assumption comes from the principle of human dignity, which is established in the Universal Declaration of Human Rights, and better defined when adopted by the International Labour Organisation (ILO) in the convention held in Geneva in 2008, which presupposes the State's guarantee to individuals of a balanced environment and the right to decent and sustainable work (SARLET, 2009).

Finally, the fourth aspect relates to the ability to infer actions and visions of an international nature, i.e. in a holistic way with the globalised community, rather than one-off, watertight solutions. It advocates the need for integration and overcoming sovereignty and national interests in favour of humanity and environmental issues as a whole (FAHRI, 2006).

Farhi ends by succinctly presenting his understanding of the political liberal environmental current and how it views, or rather understands, the relationship between economic liberalism and environmental issues:

> For de-Shalit, the environmental issue and the solution to the ecological damage caused by human activities are not a matter for the economy. The environmental issue is eminently political. Environmental policies must be the result of political criteria, decided in the public sphere, which is open to debate about the concept of the good life and its relationship to values that concern nature. This public sphere must politically discuss the concept of the good, and consider the quality of the environment as a public good, not as an individual issue, or restricted to a private group, but as a political issue that requires state intervention (FARHI, 2006, p.39).

A panorama is built that presents the environment as the locus of the relationship between individuals and society and nature, a relationship that takes place in a complex, holistic, ecological way, in its different aspects and that constitutes a legally protected asset, being in essence a social space where it is not possible to disassociate man from his relationship with the environment, be it natural or artificial.

2.4 Environmental education

2.4.1 History

The search for the concept of Environmental Education reveals the need to understand the historical process of its conception. Carvalho (2011) emphasises the growing concern of ecological movements, especially after the 1960s, with the need to raise awareness of problems related to preserving the environment.

The milestone for the conceptualisation of Environmental Education dates back to the United Nations Conference on the Environment, which took place in Stockholm, Sweden, in 1972. As a result of the need to promote environmental conservation, in the same year the UN created the United Nations Environment Programme (UNEP), which is the global authority on the subject. It is worth analysing the historical events that preceded the Conference, leading to a definition of the importance of the issue in subsequent rounds of international debate.

Against the backdrop of the polarised dispute between capitalism and communism, questions arose about the two great economic currents, above all about their ability to extend their philosophies to the rest of the world. At the same time, the struggle for civil rights, the Vietnam War and the recovery of a post-war Europe were the dominant realities on the world economic and philosophical scene, dictating the way of thinking (LAGO, 2007).

The degradation of the environment took shape during the industrial recovery of the early 1950s, when English factories experienced the **phenomenon known as "smog" (smoke + fog)**, caused by the burning of fuel without proper treatment of the residual gases. This phenomenon had a significant impact on the lives and health of workers, causing the deaths of around 8,000 people (MOURA, 2002).

Marques (2005) also points to disasters related to pollution, such as Minamata Bay in Japan and the poisoning by DDT (dichloro diphenyl tricloethane) present in the insecticides used at the time.

Concerns about environmental issues grew to the extent that it was no longer possible to disassociate natural impacts from industrial growth. In 1962, the work of biologist Rachel Carson was published under the title Silent Spring, in which the direct relationship between the use of DDT and its environmental consequences became clear.

A decade earlier, environmental political movements had already sprung up in major US cities, transforming ecological concerns into environmental issues. These influences were reflected in the decision-making of the US government, which began to propose conservationist actions and environmental policies, making them a regular item on the agenda of Presidents Kennedy, Johnson and Nixon.

Thus, Carson's denunciation of the effects of DDT use, as well as the increase in environmental concerns, drove Nixon's political agenda to establish the US Environmental Protection Agency (EPA) and ban the use of the substance (MOURA, 2002).

At the end of the 1960s, a group of scientists called the Club of Rome tackled the issue through discussion forums, **culminating in the publication of "The** Limits to **Growth"** in 1972. This study presents the relationship between population growth, increased consumption of natural resources and the resulting environmental consequences. Malthusian in nature, the analysis proved to be wrong, but it did alert states to the need to think about the environment (LAGO, 2007; BRUSEK, 1994).

Although considered erroneous by some ecologists, who seek a holistic view of the situation, the Club of Rome report is still used as an assumption for analysing the limits to growth, with the studies being used as a basis for dealing with the repercussions of the global economic crisis, especially after 2008. In this sense, Turner (2014) analyses the updated computer models and theories proposed on the basis of concepts derived from the 1972 report.

In a holistic and ecological analysis, Odum (2004) proposes the concept of Carrying Capacity, where sets of topographical, climatic and biotic characteristics influence the size of the population and its relationship with the environment. It also determines the amount of natural resources that can be extracted without jeopardising the survival of a given population. In his theory, he indicates the complexity of the factors that influence the human ecology of a given environment, considering various variables such as temperature, radiation, water and natural resources, but not in isolation from each other, but concurrently, influencing the scenario as a whole. He also points out that during the 1960s there was a collective and interdisciplinary effort to unite the vision of geographers, biologists, sociologists and other specialists in natural resources to understand human ecology and its influence on society (ODUN, 2004, p. 817).

The immediate aftermath of the nascent concern about the environmental impact of human activity gave rise to thinking about man in nature. The rudiments of a biocentric ethic emerged, based on the premise that nature has intrinsic value (UNGER, 2000).

In the same vein, Capra **(2002) defends the "ecological paradigm",** in which he proposes the development of perceptions and values of symbiotic relations between Man and Nature:

> The principles on which our future social institutions will be built must be consistent with the principles of organisation that nature has evolved to sustain the web of life. To this end, it is essential to develop a unified conceptual framework for understanding material and social structures. Sustainability does not imply the immutability of things. It is not an aesthetic state, but a dynamic process of co-evolution (CAPRA, 2000, p.17).

As a result of this change in thinking about the relationship between man and nature, the United Nations Conference on the Environment, held in 1972 in Stockholm, was attended by 113 countries, including Brazil.

As a result of the Conference, a Declaration was promulgated, laying the foundations for understanding the relationship between Man and Nature:

> According to the Declaration, man is both the work of and the builder of the environment. The two aspects of the human environment, the natural and the artificial, are essential for man's well-being and the enjoyment of his fundamental rights, including the right to life (MARQUES, 2005, p. 28).

With regard to the direct relationship of interdependence, as advocated by Capra (2001, p. **230)** "to recover our full humanity, we must recover our experience of connectivity **with the entire web of life". With the concept of the Web of Life, the author** replaces the idea contained in modern European thinking, where the world functions as a beautiful, shiny machine, with gears running at the pleasure of Man and Science in a state of deus **ex machina,** with the idea of interconnectivity and interdependence existing between the actors and their environment, or "web of life".

The Stockholm Conference, driven by the alarmist studies and publications of the Club of Rome, not only motivated the organisation of the United Nations meeting, but also dictated its environmental protectionist tone.

The Conference sought to address four main problems that had been widely highlighted: science's view of environmental issues and the changes that had taken place; the **exposure of environmental problems and crises such as the** *"mmgg" and* Minamata Bay; the disorganised densification of cities, with an even greater rural exodus; attention to harmful chemical pollutants, such as the DDT exposed by Carson (LE PRESTRE, 2000).

There was a consensus that environmental responsibility fell to humanity as a whole and not just to a particular political or economic regime. An understanding emerged that the elements were connected and that the survival of humanity as a species depended on a balanced relationship with the environment.

At the same time, discussions sought to find a common denominator for the problem, an element capable of reversing the situation and equally serving the different nations, whether their differences related to culture, development or economy. Aires (2010) points out that in all the discussions, whether to recover, minimise or preserve the source of resources, education was seen as the possible way to establish values and strategic programmes for the rescue and harmonious development of man and nature, for the development of rationality, solidarity, cooperation and responsibility towards others.

However, the economic bias was not neglected or absent. All the environmental concern that has been expressed has stemmed from the effects of changes in the structure of production following the industrial revolution. It was therefore to be expected that the guidelines would recognise the importance of the economic over the environmental. At this point,

> the declaration also shows that in developing countries, the majority of environmental problems are caused by underdevelopment, since millions of people live far below the minimum levels necessary for a dignified human existence, so developing countries must channel their efforts towards development. There is a need for industrialised countries to make an effort to reduce the gap between themselves and developing countries (MARQUES, 2005, p.29).

There was no censorship of economic development, but the recognition that there was no equality between men and therefore no conditions for all

reach the level of development and consumption found in developed countries.

For bioenvironmentalists, it was necessary to contain population growth as a measure to resolve environmental damage:

> **This group combines ideas from the physical sciences and economics to develop proposals to revamp economic models to include the notion of** ***physical limits, which involves changing our measures of "progress" andthe methods we use to promote it. Only then, these thinkers argue, can we reduce the impact of humans on the planet and prod the world towards a more sustainable global economy. Those bioenvironmentalists who focus more on overpopulation call for measures to lower population growth, like expanding family planning programmes in the Third World, and for curbs on immigration to rich countries where consumption problems are the worst. At the more extreme end, some see a world government with coercive powers as the best way to control the human lust to fill all ecological space, destroying it, often inadvertently, in the process*** (CLAPP, DAUVERGNE, 2005, p. 11).

This realisation posed new challenges, because it was necessary to work to avoid denying developing countries the possibility of economic growth, but at the same time it was imperative to maintain the consumption and economy of developed countries.

Another strand pointed not only to population growth as an element of environmental impact, but mainly to the growth of consumerism. In a global society, it was unacceptable to restrict growth and development to just one section of the population while the developed countries, historically at the forefront of consumerism, perpetuated their way of life. It was necessary to rationalise ways of life and technological development capable of meeting the needs of the population without affecting or excluding those already marginalised by the current economic system (CLAPP, DAUVERGNE, 2005).

The logical conclusion was that there would not be enough natural resources to meet the demand for levels of consumption and development that were considered high or adequate. Cultural changes were needed to bring development and sustainability into line (SACHS, 1999).

Leff (2002) pointed out that global changes in complex socio-environmental systems were necessary. These should grasp legal, social, historical, economic, ecological and cultural bases and principles in the development of society in order to favour the democratic management of natural resources. The author thus emphasised the need for an interdisciplinary approach to understanding environmental problems.

This new vision culminated in the United Nations Conference on Environment and Development, also known as ECO 92, which took place in Rio de Janeiro. It represented a milestone in international environmental policies and was attended by 178 signatory leaders.

The recognition of the environmental problem, not only in terms of its impact or man's relationship with nature, but also in terms of consumption and economic growth, was in fact the driving force behind government decisions. The Conference had the merit of acknowledging the existence of the problem and trying, within its limitations, to find and indicate solutions and changes in behaviour in order to mitigate the problems that were widely highlighted (TOZONI-REIS, 2002).

Between the Stockholm Convention and Rio de Janeiro in 1992, attention was paid to responsibility on a global scale for environmental issues, although there was little concrete action.

> **The Rio Declaration reaffirmed states' obligation to minimise extraterritorial or extra-jurisdictional environmental harm (although, as with the Stockholm Declaration, the importance of state sovereignty over its resource and environmental policies took precedence). The Declaration also called for a 'new and equitable global partnership' but in general provides little in the way of an ethic of cosmopolitan responsibility** (ELLIOT, 2002, p. 14).

ECO-92 saw the signing of important documents such as the Rio Declaration on Environment and Development, the Convention on Climate Change and the Convention on Biological Diversity, which formed the basis for Agenda 21.

ECO-92 also led to the Kyoto Protocol in 1997, which depended on the signature of the 55 countries responsible for emitting 55 per cent of the gases related to the greenhouse effect and determined, among other issues, the reform of the energy sectors, the promotion of renewable energies, the protection of green areas and the reduction of emissions of gases harmful to the environment.

With regard to the Convention on Biological Diversity (CBD), of which 150 countries have become signatories, including Brazil, Besunsan (2006) considers it to be the real milestone in the debate around global proposals, which linked economic development and sustainability.

The Convention on Biological Diversity (COP-10) held in Nagoya, Japan, established 20 targets, called the Aishi Biodiversity Targets, which must be put into practice between 2010 and 2020 by the signatory countries. The strategic goals aim to address the root causes of biodiversity loss by raising awareness among government and society; reducing direct pressures on biodiversity and promoting sustainability; safeguarding ecosystems and genetic diversity; and increasing the implementation, planning, participation and management of people's environmental knowledge.

Of all the guidelines established by ECO-92 and its consequences, such as the CBD, Agenda 21, the Kyoto Protocol and the Aichi Targets, one common item stands out: Environmental Education. This element is centralising and universal, as it was considered to be the essential tool for achieving the objectives necessary for environmental conservation. Thus, Environmental Education as a primary element of any action represents one of the most complex challenges, as it involves a complex range of concepts and interests.

2.4.2 Challenges

As a signatory to the Convention on Biological Diversity (CBD), Agenda 21, the Kyoto Protocol and the Aishi Targets, Brazil has committed itself to encouraging education and public awareness of biodiversity conservation, as well as developing programmes for this specific purpose. Article 13 of the CBD makes it clear that states have a responsibility to work on education and raising public awareness. This responsibility gave rise to the Brazilian Charter for Environmental Education, in the same year as Eco-92.

According to her, environmental education should be included on the agendas of the states, emphasising ecological concerns and the maintenance of the environment. In this way, the concepts of the environment and environmental education began to be worked on and incorporated into the routine development of scientific literacy, influencing relations between states and the way they sought to expand the means of production and consumption (LAGO, 2007).

However, there is no room here for mystification or naivety in believing that such concerns or conferences would be enough to change the economic dynamics of the world. Although various treaties and conventions were debated and signed between nations, especially the developed powers of the North, there was an exclusionary logic behind them, whereby the countries of the northern hemisphere, with their extensive and advantageous economic, technological and industrial development, decided to defend the planet's natural resources. This strategic vision, observed through the lens of Boaventura de Sousa Santos' (2011) teachings, imposes on the peripheral countries, or countries of the southern hemisphere, to develop along the lines practised by the former.

In this vein, environmental conferences and concerns do not appear as legitimate ways of safeguarding the environment and the human species, but rather as a possibility that fulfils capitalism's need to reinvent itself. Technological speculation, as well as industrial scaremongering, develops a whole new market and more areas for economic growth. The need to reinvent itself technologically, given the responsibility of anthropogenic actions on climate change, for example, was decisive for the emergence of a whole new perspective, which soon used science to justify itself and law to solidify itself (CORNETTA, 2012).

In fact, the Conference determined the development of an international environmental education programme:

> Based on this assumption, the participants in the United Nations Conference on the Human Environment, known as the Stockholm Conference, held in 1972, after discussing the role of education in the environmental issue, recommended in item 96 that authorities from different spheres of government take the necessary measures to establish an international environmental education (EE) programme with an interdisciplinary approach, aimed at both school and non-school audiences (AIRES, 2010, p.13).

This measure highlights the concern with sharing responsibilities between the various social actors, rather than just imposing the determinations of a government agenda. As a viable alternative, an interdisciplinary and international effort is proposed to raise awareness that everyone is responsible for environmental issues.

Ecological and environmental movements should strive to raise awareness, not just among governments and the means of production, but among society as a whole, where it is possible to realise that natural resources are indelibly characterised by their finiteness, as well as the difficulty of access, cost and destination, which is often irrational and socially unjust.

But what does environmental education actually mean? According to Fiorillo (2008)

> Environmental education means: a) reducing environmental costs, as the population will act as guardians of the environment; b) implementing the principle of prevention; c) establishing the idea of ecological awareness, which will always seek the use of clean technologies; d) encouraging the realisation of the principle of solidarity, in the exact sense that it will realise that the environment is unique, indivisible and of indeterminable ownership, and must be
>
> fair and distributively accessible to all; e) to realise the principle of participation, among other purposes (FIORILLO, 2008, p. 43-44).

It would be a mistake to think about the issue without the depth that an interdisciplinary analysis requires. For Morin (2007), doing science demands a certain complexity, present not only in studies, but in everyday life in general. Only by embracing complexity and interdisciplinarity is it possible to integrate and disintegrate the conceptual universe so that something new and integrated emerges. According to the author, concepts cannot be defined by their boundaries and limitations, but by their nuclei.

Morin also points out that complex thinking does not exclude simplified thinking. On the contrary, it uses simplified thinking to scrutinise concepts and build new patterns of understanding through the reunion provided by complex thinking.

With this in mind, Carvalho (2011) notes that environmental education requires scientists to think in an interdisciplinary way:

> EE fosters emotional sensitivities and cognitive capacities for reading the world from an environmental point of view. In this way, it mediates multiple understandings of the experience of individuals and social groups in their relationship with the environment. This learning process, through this reading perspective, takes place particularly through the educator's action as an interpreter of the links between society and the environment and through environmental education as a mediator in the social construction of new sensitivities and ethical attitudes towards the world (Carvalho, 2011, p. 33).

Berenice Gehlen Adams presents a series of concepts about Education

Environmental, including the Ministry of the Environment:

> Environmental education is an ongoing process in which individuals and communities become aware of their environment and acquire knowledge, values, skills, experiences and determination that enable them to act - individually and collectively - and solve present and future environmental problems (ADAMS, 2005).

Bosa & Anderlan (2011) also provide some definitions of Education

Environmental:

> The first international definition of Environmental Education was adopted by the **International Union for the Conservation of Nature** (IUCN, 1971), which emphasised the ecological aspects of conservation. The more internationally accepted Tbilisi Intergovernmental Conference (1977) defined Environmental Education as a process of recognising values and classifying concepts, aiming to develop skills and modify attitudes towards the environment, in order to understand and appreciate the interrelationships between human beings, their cultures and their biophysical environments. A
>
> Environmental education is also related to the practice of decision-making and ethics that lead to improved quality of life (BOSA & ANDERLAN, p.229).

According to Porto Gonçalves (1990), Environmental Education should be understood **as "a long and continuous process of learning a philosophy of** participatory **work** in which everyone, family, school and community, must be involved".

The Brazilian legislature has also taken care to preserve the definition of Environmental Education in law, as well as the rules for its application and development, as set out in Law 9795 of 27 April 1999 (BRASIL, 1999).

The principle is also safeguarded in the 1988 Federal Constitution, in Chapter VI, which deals exclusively with the environment. Article 225, in its entirety, lays down guidelines on the responsibility of the public entity, in its various spheres, with the environmental issue, in an integrated and integrating way together with the various segments of society (FIORILLO, 2008).

2.5 Social Representations and the Environment

Understanding the process of representations as a social, cultural and historical construction, it is necessary to conceptualise the understanding of representations, because once they are used in the sense adopted by the French historian Roger Chartier, they differ significantly from the concept adopted by social psychology, for example.

Therefore, being clear about the meaning of the notion of representations, from the perspective of cultural history, helped in understanding and looking at the documents and sources that contributed to this study. Representations should therefore be understood as the set of symbols common to a particular group or culture, in order to define and identify that group. The researcher-historian must study and understand these representations in order to understand how the world signifies and is signified, understanding the construction of thought and vision (DARNTON, 1987).

Using Cultural History means understanding the processes and transformations through the mental and historical operations that give meaning to reality. Pesavento (2003) points out that these representations take the place of reality, as a simulacrum that doesn't correspond to the perfect image as in a mirror, but which safeguards the verisimilitude between what it is and the perception of what it wants to be between the representation and what is represented.

Representations are therefore made up of multiple perceptions and discourses which, together, establish power relations and mould not only the image of reality, but the image as a result of these discourses. According to Pesavento (2003), "cultural history thus becomes a representation that rescues **representations, that endeavours to build a representation of what has already been represented"** (2003, p. 43).

At this point, the concept of representations presented so far begins to distance itself from that put forward by Chartier. For this author, representations are related to the way in which social reality is constructed, through the divisions, classifications and delimitations imposed by society in a given period and space. Therefore, the construction of representations is intrinsically related to the context in which it is inserted, whether in the political, social, economic, cultural, temporal and geographical dimensions, in order to form, from these objects, the signs that establish the representations themselves (CHARTIER, 1990).

In this way, representations also allude to the power struggles that exist within society, as elements of identity that are constructed not just by symbols and objects, but by the dynamics of conflict and power struggles that arise from them and from the various social groups in action. For both Chartier and Pesavento, representations are the ultimate elements of that social group's discourse. However, according to the first author, these discourses, at the same time as they are uttered by the social actors involved, are also producers of meaning and sense for these actors. Therefore, **"the real thus takes on a new meaning: what is real, in fact, is not" (CHARTIER, 1990, p.43).**

There is therefore no clash between truth and fiction, which is so dear to cultural historians like Darnton and Pesavento. There is, in fact, the attribution of a new meaning to what should be understood as real. For Chartier, the real does not exist in isolation from the duality between truth and fiction, but presents itself, or rather represents the reflection of one that resembles the other. The meaning of reality necessarily comes from its representation of truth, which therefore lies on the borderline between practice and representation and need not necessarily oppose each other.

Conflict and struggles are transferred from reality and the social to existing representations of this same reality. There is also no need to argue about class and social unity, since they are recognised as plural, diverse manifestations of different groups and social forces in a perpetual clash resulting from the flow of power. Therefore, the construction of a discourse, a practice and a representation are always plural and dynamic by their very nature (CHARTIER, 1990).

It is important to point out that representations are social reflections of the object that Cultural History intends to analyse and, together with practices, they create an intricate relationship that constructs meaning and gives sense to the world, at the same time as understanding it. So far we have seen the importance of recognising how dynamic this process is, never static, an inherent element of representations, which demonstrates the need for construction based on the various contextual objects that link such conflicts (CHARTIER, 1990).

However, the concept of representation established by Chartier comes up against questions about the reliability of representation in relation to the reality from which it derives. Cultural relativism emerges as a major criticism of the theories presented, and the reflections of Ginzburg (2001), another historian, who, by countering certain conditions of representations, ends up complementing the concept. The author points out that the historical narrative is made up of evidence, of a probative ballast that gives it the character of truth. Without this evidence, history becomes a mere fictional narrative, unrelated to reality and, even though it is not the truth itself, it represents it.

So, while Chartier points to representations as the duality between the presence/absence of reality or fact, Ginzburg adds that representation is the replacement of the absent thing by a simulacrum composed and formed on the basis of present and real reflections and echoes. Thus, through representation, it is possible to know and recognise reality, even if it is already absent, given its symbolic and representative composition.

Chartier's epistemological construction, based on the sociological studies of Pierre Bourdieu, is positioned precisely in the dichotomy between subject and structure. For Bourdieu (1998), currents of thought have decisive ways of understanding culture and representations. According to the author, it can be divided into two main currents, the first based on Kant and Durkheim and the second on Marx and Weber.

For Kant, culture is an exercise in creative freedom, an instrument for constructing and perceiving the world, both in the symbolic forms of Cassirer and in the social forms of Durkheim. From this current of thought, Bourdieu understands the symbolic factor and force of which, through the social function of this power of cultural goods, **he establishes "logical conformism" (BOURDIEU, 1998).**

The concept of capital used refers to the expression of value, be it cultural, symbolic or economic, and precisely has the function of attributing value to elements in order to gain an increase or decrease in hierarchy or importance within the field of dispute. In this case, we see a mixture of capitals being negotiated and disputed, including economic, symbolic and political capital. These capitals serve not only to determine the winners in disputes within the field, but also to verify the possibility of appropriating natural resources and legitimising environmental discourses, which are at the heart of symbolic disputes (ACSELRAD, 2004).

In the same vein, Moscovici's (1978) studies come close, as he disagrees with Durkheim's theory by indicating that representations are dynamic and changeable according to the group they are intended to mirror. The **mutatis mutandis** characteristic of representations reflects the daily social constructions and reconstructions of existing social dynamics.

On the formation of a group's identity, Chartier raises an important question when he asks:

> Is it necessary to identify as symbols and consider as symbolic all the signs, acts or objects, all the intellectual figures or collective representations thanks to which groups provide a conceptual organisation to the social or natural world, thus constructing their apprehended and communicated reality? (CHARTIER, 1990, p.19)

Thus, the maximum extension given to the concept of symbol, to which all forms or signs refer, **constitutes "reality". In this way,** representations are decisive for understanding the structuring structures that govern society, as they greatly determine the way in which social relations and practices occur and are embodied.

> As the matrices of differentiated discourses and practices - even the highest collective representations only have an existence, that is, they are only truly so from the moment in which they command acts that have as their objective the constriction of the social world, and as such the contradictory definition of identities - both those of others and their own (CHARTIER, 1990, p.18).

There is a close correlation between representations, the discourses emanating from them and the social position of those who use them, which under no circumstances can be considered neutral. There is a need for a complex investigation that involves, in addition to the content, the intentionality and position of the person professing the discourse being analysed.

The essence of political and legal discourses is a set of values attributed and charged with intentionality, with approaches that favour some to the detriment of others in a clear exercise of power, or symbolic violence. This consists of the monopoly of the dominant group, which forges arguments and discourses according to its interests and thus constructs and reinforces its identity (CHARTIER, 1990).

> The collective representations that incorporate the divisions of the social world into individuals and organise the schemes of perception from which they classify, judge and act; on the other hand, the forms of display and stylisation of the identity they want to see recognised; finally, the delegation to representatives (particular individuals, institutions, abstract bodies) of the coherence and stability of the identity thus affirmed (CHARTIER, 2002, p.11).

Through Marx's and Weber's conceptions, Bourdieu appropriates the concept of instruments of domination which, combined with Kantinian and Durkheinian understanding, lays the foundations for the conception of symbolic power and the economy of symbolic exchanges (BOURDIEU, 1998).

Through the concept of Field, Bourdieu presents the locus of social action where power relations are manifested, the place where the dispute between social actors over specific interests takes place. It is the place where society meets to dispute interests.

> The assumption of the notion of field in the approach to a given research object leads to the necessary consideration that it does not exist in isolation and is not isolated from a set of relationships from which it draws the essential properties. (BOURDIEU 2007, p. 27 apud SILVA, 2012, p. 29).

Focussing on the environmental issue, the concept of field defines the environment as a social space of domination and conflicts not unlike other fields, such as politics and education. As you might expect, these conflicts and disputes are governed and organised by their own rules, which take into account geographical space, the actors involved and social capital, among other elements. Notably, the environmental field breaks with the idea of a static consensus related to the environment, as it is a place where disputes over power and dominance are constant, both in defence of visions and material and symbolic issues (SILVA, 2012).

In this study, understanding Bourdieu's concept is necessary to overcome the dichotomies of a functionalism of symbolic productions. In fact, the symbolic systems argued by the author have political functions of legitimising and dominating groups, both internally and externally, which corresponds to symbolic violence per se. However, there is also the character of identity recognition linked to the symbolic question, as well as domination - forming the basis of what he recognises as a structuring structure (BOURDIEU, 2002).

The concept of habitus refers to the question of a structure that is structured to function as a structuring structure, in other words, a structure that ends up serving to guarantee the maintenance of representations, being unconscious and at the same time belonging to the conscious of the collective that struggles to sustain it. Thus, behavioural patterns, when they are solidified and understood as official patterns, make it possible to perpetuate practices within representations and are supported by the concept of habitus, and indicate the real principles of socially accepted and adopted practices, as defined by Bourdieu:

> A system of durable and transposable dispositions which, by integrating all past experiences, functions at every moment as a matrix of perceptions, appreciations and actions - and makes it possible to carry out infinitely different tasks, thanks to the analogue transfers of schemes (BOURDIEU, 2002, p. 21).

Habitus plays an important role within the environmental field, as it concentrates the strategies that are in dispute in environmental conflicts and the behaviour of each element according to their position within the structure of the field. The disputes over power and symbolic capital, whether institutionalised or not, occur through the mediations of habitus, which determine whether the status quo is maintained or altered. The symbolic capital referred to here is defined by Bourdieu (2002) as the idea of power, which determines the hierarchical position of each element within the field and the way in which representations are incorporated and negotiated.

Chartier's concept of representation indicates the discourse and conceptions of certain trends, the appropriation of reality constructed and negotiated by these segments, thus differentiating between them. It serves as a simulacrum of reality in that it tries to reflect its perception while not allowing itself to replace it, but only to represent it as faithfully as possible to the discourse of those who mimic and invoke it.

Aires (2010), in adopting the concept of social representations in his study and working with representations of Environmental Education, used the Theory of Social Representations defended by

Santos and Almeida and by Moscovici. In this way, she sought a scientific understanding of the common sense and everyday perceptions that exist in a given group, verifying the various manifestations, organisations and expressions of these actors through their discourses. It also demonstrates the simultaneous role of discourse as a structure and as a structuring element of the social realities represented:

> As such, Moscovici's study aimed to demonstrate, in the scientific field, that individuals were endowed with capacities that could modify society, being active and creative subjects who construct reality and reconstruct themselves as human beings in that same reality, through relationships that are created in a process of elaboration, construction and appropriation of objective reality. Hence the complexity of the term **"Social Representations"** (AIRES, 2010, p.21).

However, it is important to emphasise that Aires uses as a basis the theories of social representations in psychology, constructed by French psychologist Serge Moscovici in 1961. Mosvovici's (2007) studies present these social representations as a set of explanations through the interactions of everyday life, constituted by the subject's mental work on their perception of reality, formed from the image of the object. They are also a mental effort to bring together what was far away and make recognisable what was strange.

Aires (2010) also quotes Santos and Almeida's understanding of social representations:

> [...] to refer to the knowledge produced in common sense. But not to all knowledge, but to a form of shared, articulated knowledge that constitutes a lay theory about certain social objects is to refer to a theoretical model, a scientific knowledge that aims to understand and explain the construction of this lay knowledge, of these common sense theories (SANTOS and ALMEIDA apud AIRES, 2010, p.21).

The importance of understanding representations converges in Aires's thesis and in the present study when Santos and Almeida look together at the meaning they give to the study of the phenomena of these representations:

> For Santos and Almeida (2005), studying the phenomena of Social Representations means understanding the word that was said, the gesture that was expressed, the attitude taken towards certain contexts and situations that surprise individuals, groups and society. Wherever there are speeches, written or spoken messages, media images in the crystallised behaviour of human beings and in material and spatial arrangements, representations are present (AIRES, 2010, p.21).

Studying discourse in order to understand the elements that make up these representations is indelible for a correct understanding of what is said and what is practised. This leads to the objective of establishing the relationship between praxis and representations, or in other words, the dichotomy and similarities between discourse and practice (RANGEL, 1999).

Still on the subject of social representations and the environment, the discussion involving the pursuit of sustainability as one of the key aspects to be achieved through environmental education is brought to light. To this end, the studies of Ignacy Sachs (2002) are used to help conceptualise the various types of sustainability.

For Sachs, sustainability comes in various forms, with different connotations and approaches, or a diversity of perspectives on the same object. Sustainability can therefore be ecological when it seeks to maintain and balance natural resources linked to the production chain; environmental when it refers to the resilience of the environment in the face of anthropic influences and damage; and social, political and economic when it relates to human and population issues, the process of perception and citizenship, as well as the production chain and the way in which the relationship between man and the environment is restructured and reinvented (SACHS, 2002).

The great challenge lies in conceptualising or even achieving sustainability in such a way as to guarantee the provision of the needs of the current generation without compromising the environment's capacity to provide sustenance for future generations, in a tangle of causes and consequences that is very close to Sachs and Capra's understanding.

In the conceptualisation of the environment in relation to representations, Paula Bruger, when relating social groups and their relationship with the environment and their understanding, says:

> There is a broad consensus that the concept of the environment should encompass a totality that includes natural aspects and those resulting from human activities, thus being the result of the interaction of biological, social, physical, economic and cultural factors (BRUGER, 1996, p.53).

The author strives to conceptualise the environment beyond its technical or natural dimension, adding the subject and the historical-social dimension, using representations. Bruger, in the same vein, conceptualises development and sustainability through a Marxist conception. For her

> sustainable development (...) must be one that invokes a new ethic, a redefinition of what material and spiritual well-being is, in the interests of the majority of the population, reversing the present state of degradation of life. In this new ethic, the hegemonic concepts of the environment, science, technology and education (encompassing all the ways in which knowledge is formed) must undergo a profound epistemological revision, since they are, in the current framework, inextricably associated with the causes of the degradation of life, to the extent that they ideologically and materially underpin the dominant system of production (BRUGER, 1996, p.74-75).

Complementing Bruger's idea, Morga's view is that the process of education and sustainable development needs to open up dialogue with historical and cultural perspectives so that the various groups and representations can participate, not as peaceful agents, but as actors who understand their responsibility for the needs of the future:

> It is a process of change in which the exploitation of resources and the orientation of investments, the direction of environmental development and institutional change must take into account the needs of future generations (MORGA, 1996, p.04).

Lefebvre points to representations as one of the perspectives for understanding the development and dissemination of concepts (statements, explanations) and images of the world.

"reality", as subjects perceive and construct it. Another important factor considered when studying representations from Henri Lefebvre's (1983) perspective is that they interfere in social practice while interpreting it.

Therefore, education, whether environmental or not, is an indelible element of practice for representations. For Tardif, there are three conceptions of educational practices.

The first is that which understands education as an art, linked to the Greek tradition, according to which the educator is not a scientist, because their objective is not to know, but to act and form, in the specific context of the situation, concrete human beings. The educator's job can be correlated with that of a craftsman, working and moulding the student into a predefined format (TARDIF, 2002).

The second concept sees education as a technique guided by values, and again we can see the Hellenic heritage in this way of educating, although it is more in keeping with the spirit of Modernity. This, in turn, has at its core the moral dichotomy between the objective, as in positivism, and the subjective, the existential search, as well as the dichotomy of God versus Nature and God versus Science (TARDIF, 2002).

Finally, the third concept of education presented by Tardif refers to the concept of interaction, and is supported by hermeneutics, symbolism and interaction through communication. This model forces interaction, deconstruction and construction of paradigms and collective representations, through the students' experience and relationship with the environment, as well as the exchange of their experiences and conceptions, attributing meanings and interacting in a normal and natural way and not as a solemn socio-historical event like the previous concept (TARDIF, 2002).

When explaining and conceptualising the environment and environmental education, Reigota states

> a determined and/or perceived place where there are dynamic and constantly interacting relationships with natural and social aspects. These relationships lead to processes of cultural and technological creation and historical and political processes of transformation of nature and society (REIGOTA, 1994, p.21).

It can be concluded that the representations present in Environmental Education link the joint and interdisciplinary understanding with the education of ethical and political values, which seeks the transformation of social paradigms and their relationship with nature, an indelible theme for the survival and maintenance of the relationship, given the already verified interdependence of the actors (REIGOTA, 1994).

2.6 Urban Space

Urbanisation is a concept worked on by both the social sciences and the natural sciences. The environmental problems of urban growth can be seen in the works of Castell (2009), where urbanisation appears as a recurring problem. The author points to the existence of two currents, one dealing with the spatial concentration of the population and the other with the diffusion of values, attitudes and behaviours of urban culture, which is the focus of this study.

When discussing the urban environment and man's action in transforming it, Pacheco (1992) points to the need for integrated management between natural and social space, seeking ecological and sustainable balance for urban populations.

By thinking about urban space as a locus of coexistence and social development through the lens of Sociology, Geography, History and Law, a dialogue is built between different understandings that allows points of intersection to be located between their conceptions, where considerations about the social function of urban space can be found.

Correa (1989) sees urban space as a set of juxtaposed realities, which can be fragmented and articulated at the same time.

> The space of a large capitalist city is, at first glance, the set of different land uses juxtaposed to each other. These uses define areas such as the city centre, where commercial, service and management activities are concentrated, industrial areas, residential areas that are distinct in terms of form and social content, leisure areas and, among others, areas set aside for future expansion. This complex set of land uses is, in reality, the spatial organisation of the city or, simply, urban space, which thus appears as fragmented space (CORREA, 1989, p.14).

The author also complements the issue of urban space by pointing out several of its characteristics, such as a fragmented and articulated space, a reflection of the social condition, and defining it as a symbolic field of struggles.

In line with this thinking, Lefebvre (1999) places the city as a space of commercial flourishing of paramount importance after the first Industrial Revolution, observing that the configuration of urban space was linked to capitalist development and, in its second phase of constitution, was related to the emergence of industries, transforming it into the space of industrial cities, going beyond the sphere of commercial relations.

Cosgrove (2003), in turn, points to urban space as being a space of human construction, a social construction that can be perceived through its symbolic codes and values, which are the fruit of the culture of that specific time and place.

Aligned with the Marxist bias, it demonstrates the importance of the economic perspective in understanding and constructing the concept of urban space, where it is constituted in the organisation of the mode of production around the city. In this way, it can be seen that the space of economic production is also a space of cultural and symbolic production, and this discussion opens up space for sociological culturalism.

For Bourdieu (1989), social actors seek domination and appropriation of these urban spaces through the dispute over symbolic capital, which he divides into three types - cultural capital, relating to the production of formal or informal, erudite or popular knowledge; social capital, relating to social relationships, friendships and contacts in favour of some benefit, whether real or symbolic; and economic capital, relating to the accumulation of goods and material resources available to social actors. For the author, urban space is a privileged locus of urban social disputes, where there is a constant struggle for symbolic power between different groups aimed at maintaining or changing the established social order.

Urban space is the stage for these disputes, which do not occur in a disorderly fashion, but serve to reaffirm and maintain existing symbolic power, fuelling the rules of power and serving as an extension of the individual. Historically, it has been a locus where the inclusion or exclusion of certain social groups has been promoted, with the dispute being the element that structures contemporary society, at the same time as this society structures the dispute, using the concept of structuring structure.

When it comes to disputes over power, and understanding urban space as a place where contemporary society struggles in this sense, Foucault (1979) points to a place where this force actually acts, disciplining the social structure through the control of these spaces, their classification and political, spatial and social organisation.

The legal doctrinal understanding of urban space contrasts this concept with what is rural and related to the countryside, defining this term as Artificial Environment, or even all those places where there has been human manifestation, which is reflected in the alteration of the natural environment, calling it urban territorial space. In this understanding:

> [...] the artificial environment is comprised of the built urban space, consisting of the set of buildings (called closed urban space), and public facilities (open urban space). In this way, all built space as well as all spaces inhabitable by human beings make up the artificial environment (FIORILLO, 2008, p. 300).

But how does law, as a science, differentiate between urban and rural areas? The rural perimeter, due to the sparse nature of the buildings, defines the exclusion of the rural from the urban.

There is also express legal provision for urban space as an artificial environment, as set out in articles 182 and 183 of the 1988 Federal Constitution, as well as in the City Statute, Federal Law No. 10,257 of 2001, which provides for public order and social interest behaviours related to and defined as urban.

The brief conceptual overview above leads to the construction of the perception of urban space. In history, the constitution of urban space is understood as a long process that began with the first settlements and the first civilisations around 8,000 years ago.

Thus, it has to be considered that the modern, or even post-modern, urban space differs greatly from that of pre-modernity, rescuing its differentiation from the rural space when it inverted the logic of domination and importance, consistent with the emergence of capitalism and the changes in the industrial mode of production.

It was in the Middle Ages that work began to specialise and urban space began to determine and monopolise the importance of the countryside. The rebirth of urban centres after the fall of a weakened Roman Empire, as well as the emergence of new means of production, dictated the establishment of the relationship and understanding of this urban space.

The history of the city and urban space is therefore confused, especially from a Marxist perspective, with the history of the specialisation of labour. In this sense, we turn to Mumford (1998) where:

> [...] it was in the monastery that the ideal aims of the medieval city were put in order, kept alive and ultimately renewed. It was also there that the practical value of restraint, order, regularity, honesty and inner discipline was established, before these qualities were passed on to the medieval city and post-medieval capitalism in the form of inventions and business practices: the clock, the ledger, the ordered day (Mumford, 1998, p. 41).

The specialisation and abstraction of work, combined with the process of acceleration in the industrial period of the 19th century, established the utilitarian character of the factory that took over urban space, thus understood not merely as a space for man's environmental transformation, but also as the space of chimneys and factories. It was the victory of utilitarianism over humanity, which equated urban space with economic production space (BENÉVOLO, 1994).

This means that urban space is the artificial environment, dominated and altered by the intentional and conscious action of man, whose value is reflected in the utilitarianism of the means of production. It is also the privileged locus for the exchange of capital, be it economic, social or even symbolic, thus reflecting the inclusion or exclusion of groups in the constant alternation and reproduction of the means of control and power.

At this point, it is important to note that the notions of public and private are often confused or do not respect each other's boundaries, as can be seen in historical materialism, especially in the description of the process of enclosure of English communal lands described by Marx (1996).

For Gomes (2002), the understanding of public space as one with free access is limiting and incapable of encompassing the various dimensions it contains. Taking as an example the case of universities, hospitals and other offices which, although public places, have restricted access to authorised groups, the author also points out that public space often takes on the characteristics of private space, i.e. it doesn't necessarily include free access, and its use can be private or collective (GOMES, 2002).

Hertzberger (1999) establishes differences between these two spheres, using the criteria of access, forms of use and who these spaces are for to define them. This is complemented by DaMatta

(1997), who establishes a symbiosis between these two universes, comparing, from his perspective, the home and the street, in a relationship in which there is a dynamic dependency between the two spheres. Santos (1985) points out the need to consider not only the fixed elements that exist in them, but also the flows that operate there.

Although from an economic point of view about the constitution of urban space, especially public space, as the place where society's productive forces are established, Santos (1985) differentiates fixed elements as those that come from long ago and are consolidated in the urban fabric, while flows are, in a way, derived from and dependent on the former, and may or may not become fixed at a certain moment in their history.

For Correa (1993), urban space is both fragmented and articulated. It consists of capitalist space and as such brings with it all the inequality inherent in this system of production. Space is a reflection both of the actions that take place in the present and those that took place in the past. Thus, a concept of public space can be seen as a place of conflict and social confrontation, and not just a geographical space for common or collective use.

> Strongly associated with monetary income levels, the fragmentation of urban space and its full awareness lead to social conflicts such as workers' strikes and particularly the so-called urban social movements. The space of the city is thus also the setting and object of social struggles, since these are ultimately aimed at the right to the city, to full and equal citizenship for all. Urban space thus becomes a field of struggle (CORREA, 1993, p.17).

In defining public and private space, Lefebvre determines three levels of characterisation: a global level, an intermediate level and the private level. The global level is where the essential relations with the capital market and space planning policy are found, where there is interference in the practical-sensible and the immediate, where public buildings and constructions are built. The intermediate level is related to the city, where commercial structures, transport and other urban services are located. It is called intermediate because it is precisely a level between the sphere of global and macro relations, and micro relations, or those relating to the private sphere. Finally, the sphere of the private level is where the buildings for housing, camps, houses and slums - machines for inhabiting - are located (LEFEBVRE, 1999).

The private space contrasts with the public space because it is the place of daily and family life, where the intimate relationships of individuals are limited, circumscribed in Lefebvre's machines of inhabitation. This is also the understanding of the Brazilian legal system, which further limits the private sphere to what concerns only the life of the individual and their family unit, while not interfering with the rights of others, such as neighbourhood rights.

In contemporary times, the process of public spaces being taken over by private spheres continues, especially when we understand public urban space as a place where power is disputed and maintained. In understanding these disputes, Foucault (2001) points to a type of spatial sacralisation that prevents space from being recognised as public, giving it exaggerated protection in the private sphere, which is incapable of combating this takeover:

> And perhaps our lives are still commanded by a certain number of oppositions that cannot be touched, which institutions and practices have not yet dared to attack: oppositions that we accept as entirely given: for example, between private space and public space, between family space and social space, between cultural space and useful space, between leisure space and work space; all are still driven by a secret sacralisation (Foucault, 2001, p. 131).

The 1988 Constitution addressed issues relating to the use and occupation of urban land. Meirelles (1993), referring to the Athens Charter, lists four primary functions for cities: housing, work, circulation and leisure. To this end, he points to the document drawn up during the International Congress of Modern Architecture, held in the 1930s in the Greek capital, where these primary functions were recognised and attributed as necessary characteristics for urban space, which consolidated the understanding of the modern city.

However, the Athens Charter should not be seen as a limiting factor, but rather as an element of cohesion and influence in the Brazilian constitutional understanding of urban space, a topic that will be better addressed when the Palmas Master Plan is discussed.

The social function, in this sense, strives to guarantee family and individual housing, the development of economic activities capable of providing sustenance and dignity, the free movement of inhabitants and quality of life - key elements for Brazilian legal understanding (GARCIAS & BERNARDI, 2008).

Having analysed the above, we can once again see the importance of the economic sphere in determining the social use of space, whether urban or rural. The activities that take place there are the result of a construction over time, but they should not be limited to this sphere for their perception and economic-political development, as Foucault rightly points out:

> We need to follow up on this and not just say that space pre-determines a history that in turn modifies it and is sedimented in it. Spatial fixation is an economic-political form that must be studied in detail (FOUCAULT, 1998, p. 212).

Urban space is a construct, a human artefact, a space transformed by and to meet the expectations and uses that society assigns to it. It's certainly a complex historical process, but that's not all. The very choice of location already gives indications of what that urban space that is being consolidated will be like and how it was conceived.

Nor can the study be limited to a purely economic perception, otherwise, as Souza points out, we will be limited to a single vision:

> Unfortunately, in Foucault's case, as in the case of most Marxists, the interest in a social critique that strongly takes into account the spatial dimension of society was much more about analysing

> and denouncing heteronomous social relations and spatialities than about investigating autonomous social relations and spatialities, or at least those that led to substantial gains in autonomy (SOUZA, 2006, p. 316).

Social agents and their relationship with urban space also influence these dynamics. They are the owners of the means of production, landowners, property developers, the state and excluded social groups, all acting in their own interests in a constant dispute to occupy the urban fabric or even prevent the occupation of this fabric by those who will not bring the desired profits from the use of the land (CORREA, 1996).

To think of the discourse of social welfare and the guarantee of human and individual rights contained in the Constitution is to disregard all the existing and active forces in the formation of urban space. Although the right to the city is provided for in the Constitution, and its ownership and property must fulfil a social use that is foreseen or expected, there has been little effectiveness in this regard.

Brazil has a Constitution that protects human rights and dignity. However, the rest of the legal system is concerned with property, a remnant and rancour of a Civil Code from before 1946. Even today, with a new Civil Code, the protection of property takes precedence in national judgements, guaranteeing the maintenance of power and the status quo.

The urban space is the locus of interaction between these different groups, a privileged place where freedom of interaction must be respected, seeking their inclusion, contrary to the logic of capitalism, which needs the excluded as one of its perpetuation mechanisms. Understood as a place of confrontation, where the excluded and other social actors come together, urban space is a locus, in other words, the place where symbolic exchanges take place, perpetuating the dynamics needed to influence the interpretation of cities and their inhabitants (BOURDIEU, 1989).

In this way, the Law, through its normative systems, is given a prominent role as a guarantor of the dynamic of assimilation of the heterogeneities and differences that exist between the city's participants.

The new Athens Charter, drawn up in 2003, emphasises the social function of urban space:

> to conserve the cultural richness and diversity built up over history; to connect through a variety of functional networks; to maintain a fruitful competitiveness while striving for collaboration and co-operation; and to contribute to the well-being of its inhabitants and users (NEW ATHENS LETTER, 2003).

When considering the Statute of Cities, the Brazilian legislator, bearing in mind the social character of urban space, determined not only urban functions, but also social and management functions for these spaces. Thus, to the four pillars of the first Athens Charter - housing, work, mobility and leisure - were added the functions of education, health, safety, protection, provision of services, planning, cultural preservation and urban sustainability. Thus, the social function of urban space now has three dimensions, or three axes, each with its own specific determinants (BRASIL,

2002).

Establishing a common urban space meant establishing a stage where various social actors found a field of action, where structures were structured at the same time as they were structured in disputes over power and resources. Thus, urban space presents itself as a place of both social inclusion and exclusion. Its various agents act to maximise their gains in relation to property, while denying access to those who can offer nothing in economic terms. However, the very means of exclusion ends up including these groups in the urban context, in a different way to that intended, but still indelible to its conception.

In relation to the public and the private, it has historically been noted that private interests have overlapped with public interests, even though there has recently been legal provision for the opposite. It is precisely in this context that the concept of the social function of urban space and the social function of property arose, in the wake of the fundamental rights provided for and protected by the constitution.

The social function of urban space is a concept that seeks to balance human and power relations within the city. In this context, public authorities must use this principle to establish not only the search for, but also the guarantee of this desired balance.

2.7 Environment and Law

The essence of law is a constant within the diversity and plurality of discourses, a synthesis of the will of society in the construction of its order. According to Lyra (2000), this essence

> [...] arises from the constant, in diversity, and is technically called ontology. It should only be emphasised that a dialectical ontology, as indicated by the Hungarian philosopher Lukács, is based on phenomena and it is from **these that it seeks to deduce the "being" of something, thus searching** within the chain of transformations itself (LYRA, 2000, p.30).

For Kelsen (2009), law represents the monopoly of social force, which acts to guarantee peace through a cohesive legal order that meets the needs and desires of society. According to Kelsen, the state represents the manifest interests of this society and, therefore, its "social peace" and established order, which originate from the collective will and for the collective will.

It is widely believed that environmental law is a recent subject, which emerged in the mid-20th century, partly as a result of neo-Malthusian alarmism. However, the principles and goods protected by this branch of the law have their roots in antiquity itself, from the Code of Hammurabi to Roman law, the cornerstones of Western law (MARUM, 2002).

Even in antiquity, there was a concern, if not with preserving the environment, with maintaining the conditions necessary for human life. It was therefore to be expected that the codices throughout the ages would record and instruct on the use of existing forests, rivers, lands and natural resources.

Environmental Law in Brazil, which has been around since the Manueline compilations of 1514 (GARCIA, 2010), was only separated from Civil Law recently, when it gained autonomy due to its growing importance. In this way, it even stood out in terms of its principles, to the point of being recognised autonomously with the advent of the 1988 Citizen's Constitution. According to Fiorillo (2008),

> Environmental Law is a new but autonomous science. This independence is guaranteed because environmental law has its own guiding principles, set out in article 225 of the Federal Constitution (FIORILLO, 2008, p. 26).

Therefore, when it comes to environmental law, it is based on the rights of the community, i.e. it is in the interest of society as a whole that it be enacted and observed. According to the jurist Piva (2000), environmental law requires the involvement of both the public authorities and society. According to the author, there is a need for the state to intervene in order to guarantee natural assets, acting both as an enforcer of the law and through the principles of protection and precaution. Piva also points to the need for the state to provide environmental education in order to make society, both individuals and companies, aware of their responsibility to preserve the environment.

This is also the view of Rangel, who states:

> In line with the guidelines that emanate from this dogma, Environmental Law is extended to everyone, and is therefore a public interest. Given these premises, it is possible to assert that everyone has the right to enjoy and the obligation to respect the environment, and that any individual has the defence and prerogative to enjoy it in particular. As such, the state, with a view to collective protection, must act through its institutes, including in a coercive manner, with the aim of ensuring quality of life for all (RANGEL, 2012).

Although the author argues that it is the state's duty to act, there is not necessarily unanimity in doctrine about the limits of this participation or act, varying according to the doctrinaire, the jurist and the case in question.

However, there is a consensus in the doctrine regarding the principle of the supremacy of the public interest in protecting the environment over private interests. In this way, the need for action by the Public Authorities in their different spheres is evident. Rangel (2012) adds that it is up to the Public Authorities to act not as owners of the environment, but as administrators, managing and respecting the rules of Environmental Law.

> In this way, the public authorities are not the owners of environmental goods; on the contrary, they must act as administrators of the res omnium, managing it and respecting the rules of administrative law, such as efficiency and accountability (RANGEL, 2012).

As seen above, there is a concern to demonstrate Environmental Law as a protection of typically diffuse goods, i.e. those that belong to the community, and therefore a primary need to understand everyone's responsibility and participation in both the process of protecting and promoting a sustainable environment.

This lists the fundamental principles that guide environmental law: the polluter pays principle; the prevention principle; the participation principle; the environmental information principle; the environmental education principle; and the ubiquity principle.

In the understanding of Environmental Law, precaution is more important than reparation or punishment, since in many cases the damage caused tends to be irreparable. This leads to the understanding that prevention and Environmental Education are key processes for the effective protection of the environmental good. In this sense, Padilha (2010) states that

> Environmental law is concerned with risk and not just damage and, in this sense, is based on the pillars of prevention, since any decisions involving the environment can affect the quality of life of present and future generations and of complex and interrelated ecological systems, and should therefore not be guided by haste, hastiness or irresponsible attitudes, which are not in line with the care and respect that should guide the goal of preserving life in all its forms (PADILHA, 2010, p.253).).

In a nutshell, the polluter pays principle is understood as the legal imposition of the duty to make reparation to those who have caused damage to the collective natural heritage. It is laid down in art. 4 of Law 6.938/81 and also in the 1988 Federal Constitution, which in its art. 225 establishes that

> conduct and activities considered harmful to the environment will subject offenders, whether individuals or companies, to criminal and administrative sanctions, regardless of the obligation to repair the damage caused (BRAZIL, 1988).

It is important to point out that the principle does not encourage or merely seek compensation for environmental offences and reparation for the damage committed, but rather has at its core the similarity found in criminal law. This is intended to curb the offence and, once it has occurred, it has a pedagogical nature so that the offender no longer commits the offence and the damage perpetrated.

Likewise, the understanding of the principle of prevention, also known as precaution, is supported by the same legal provision. This complements the intention of the Brazilian legislator when it determines and equates the risk with the damage actually caused to the environment.

As for the principle of participation, it has the power to establish a universalisation of those responsible for protecting the environment, calling on various social actors to take responsibility and not just the entities of the federative pact. As this is a good inherent to everyone, it is to be hoped that society as a whole will be involved in responsibility for implementing public policies related to the environment and sustainability. This requires sharing responsibilities with all the players who use and are served by it (PADILHA, 2010).

Two primacies derive from the principle of participation: that of environmental information, which states that all environmental information necessary for the formation of awareness, and that of environmental citizenship must be widely disseminated. This understanding is in line with art. 220 of the 1988 Federal Constitution, which deals with the free circulation of information and social manifestation. Likewise, Environmental Education is a pillar of the Law, included in the 1988 Magna Carta, and its aim is to raise awareness of environmental preservation. It received further attention and better formatting with the National Environmental Education Policy, through Law No. 9.795 of 1999, which determines the defence and education of the environment as one of the responsibilities of the state and society.

Finally, the principle of ubiquity determines precisely the diffuse nature of the legally protected good that is intended to be protected. It thus determines that the environment is universal, belongs to everyone and is present everywhere, and must always be taken into consideration and protected by any and all human activity.

Fiorillo (2008) ponders on this principle:

> In this way, it can be seen that environmental law requires not only **global "thinking", but also** local action, because only in this way will it be possible to act on the cause of environmental degradation and not simply on its effect. In fact, it is necessary to combat the causes of environmental damage, and never just the symptoms, because if only the symptoms are avoided, the conservation of natural resources will be incomplete and partial (FIORILLO, 2008, p. 46).

Elevated to the category of a protected good by the Constitution, there are a series of mechanisms aimed at protecting it, as well as guaranteeing state supervision in related matters. According to Kamphorst (2014), the philosophical tradition linked to the liberal theorist of justice, John Rawls, is not opposed to state intervention, despite defending positions of neutrality. The author points to the same understanding as Rawls, according to which the state is called upon to intervene in the supply and guarantee of equity in the distribution of public goods, and it should be up to each society to establish which goods should be included in this category (RAWLS *apud* KAMPHORST, 2014).

In Brazil, environmental principles were elevated to Constitutional status, which guaranteed them deference in relation to other areas of law. Although it was born in the midst of the civil sphere, it acquired an autonomous aspect with the 1988 Constitution, where its normative concepts were related to environmental issues, not only natural ones, but also artificial and social ones.

Article 255 of the Federal Constitution states that

> Everyone has the right to an ecologically balanced environment, which is a good for the common use of the people and essential to a healthy quality of life, imposing on the public authorities and the community the duty to defend and preserve it for present and future generations (BRASIL, 1988).

What stands out here are the principles already listed above, set out in the article's heading, which makes it possible to see their implicit provision in the legal text. Its autonomy derives from this constitutional rule, which asserts environmental protection and places it among the guiding principles of Environmental Law (FIORILLO, 2008).

As far as legal protection for the environment is concerned, it finds its greatest adversary in the discourse of economic development. Although the rhetoric seems worn out and outdated, it still finds support in the development discourse linked to sectors of the economy. A commonplace in Brazilian discourse, both economic and social development must be seen through the prism of sustainable development (PADILHA, 2010).

In the 1988 Constitution, there is also express provision for the development of the economic order, listed as a foundation of the Brazilian state, in art. 170. However, this same provision, in its item VI, added by Constitutional Amendment 42 of 19 December 2003, establishes that the protection of the environment must occur through differentiated treatment when there is an impact from the economic activity carried out (BRASIL, 1988).

There is nothing to argue about the clash of constitutional principles between environmental preservation and economic development. The legislator's will to determine a higher hierarchy for the principle of environmental protection has been substantiated, with the aim of safeguarding development, as long as it is sustainable (PADILHA, 2010).

With the constitutional provision to protect the environment, the challenge has become to find ways to make this protection effective, including the development of environmental education as a compulsory topic in school education.

Law No. 9.795 of April 1999 defined Environmental Education in its Article 1:

> Environmental education is understood as the processes through which individuals and the community build social values, knowledge, skills, attitudes and competences aimed at conserving the environment, which is a good for the common use of the people, essential to a healthy quality of life and its sustainability (BRASIL, 1999).

The same law, in its second provision, already determines that Environmental Education is a permanent and fundamental component of national Basic Education, and must be present in an articulated way in the content and in the various modalities of the educational process. According to the joint understanding with the Law of Guidelines and Bases of Education, the content of Environmental Education must be present at all levels, from Primary Education, which is the responsibility of the municipalities, to Higher Education and postgraduate studies, which are the responsibility of the federal government.

It is therefore to be expected that there will be a legal provision ratifying the importance of Environmental Education at all levels of education, whether formal or non-formal. This attribution of the state to the role not only of *amicus curiae*, but also of *custus legis* regarding the promotion and development of environmental education in society, as provided for in article 12 of Law 9.795/99 in intelligence and being complemented by Decree Law No. 4.281 of 2002.

The legislation also takes into account the current that defends the development of autonomy and responsibility of local communities in relation to the environment.

> ***These thinkers stress the need to, in the words of some, "think globally, act locally." In other words, understand the global context, while at the same time acting in ways suitable to the local context. These thinkers advocate bioregional and small-scale community development-because they firmly believe that a stronger sense of community will fulfil basic needs and enhance people's quality of life. Such development would help to reduce inequities and levels of consumption that are out of balance with the world's natural limits*** (CLAPP and DAUVERGNE, 2005, p. 16).

Brazilian legislation related to Environmental Education converges in this direction, thinking globally about environmental impacts while raising awareness for local action by communities, assigning everyone a share of responsibility for conserving the environment in an integrated and holistic way. This perception stems directly from the Rio-92 Convention, as well as the Convention on Biological Diversity, that it is necessary to engage the whole of society in order to minimise environmental damage, calling everyone to account (ELLIOT, 2002).

The body of Law 9.795/99, which made environmental education compulsory, states that it cannot be watertight, segregated from other areas of knowledge, but must integrate the knowledge to be developed in an interdisciplinary way, as can be seen in article 10 of the aforementioned law:

> Art. 10: Environmental Education will be developed as an integrated, continuous and permanent educational practice at all levels and modalities of formal education.
>
> § Paragraph 1 Environmental Education shall not be implemented as a specific subject in the teaching curriculum.
>
> § Paragraph 2 - In postgraduate and extension courses and in areas focused on the methodological aspect of Environmental Education, when necessary, the creation of a specific subject is authorised.
>
> § Paragraph 3 - In technical and professional specialisation training courses at all levels, content dealing with the environmental ethics of the professional activities to be carried out must be incorporated. (BRASIL, 1999)

There is a particular concern on the part of the legislator to ensure that there is a cross-cutting nature to the themes covered in education, in order to prevent the plastering of knowledge related to the environment. There is a legitimate concern not to format the teaching plan related to Environmental Education in a Cartesian way, so that it can dialogue with society, leaving the plane of ideas and approaching the context of the students' education.

The National Education Plan, made up of Law No. 10.172 of 2001, seeks to establish precisely this cross-cutting nature advocated by the legislator's intention, defining the responsibility of joint action between educational bodies and the National Environment System - Sisnama - and the National Environment Council - Conama. According to the legal text, it is the duty of these bodies to get involved with the different segments of society, with a view to developing a prosperous network for Environmental Education.

The legal provision stems mainly from Brazil's accession to the Convention on Biological

Diversity agreed during ECO-92, which made up Agenda 21 and established the need for the Brazilian state to encourage and develop respect for and preservation of the environment.

There is, therefore, a real effort in Brazilian and international legislation to determine the correlation between the bodies and organisations responsible for promoting environmental education, with the aim of ensuring that this promotion takes place across the board, enabling and encouraging the relationship between education and the environment and other knowledge that shapes citizens.

CHAPTER 3

METHODOLOGY AND CUT-OFF

3.1 The methodological approach adopted

In order to achieve the proposed objectives, we analysed the discourses adopted by the political representations that made up the municipality of Palmas and how these representations met with practice.

A document review and analysis was carried out, in which it was necessary to conceptualise the understanding of the Environment, Environmental Education, Urban Space and the context of the Environment in relation to the Law and current national legislation. The explanatory research sought to answer the questions raised by analysing the data, the discourse, contrasting the factual situation and the documents.

Below is a description of the methodological approach adopted to be combined with the theoretical basis used and presented above.

3.1.1 Discourse Analysis in Representations

Discourse analysis was fundamental to the development of the work, using Michel Foucault, Laurence Bardin and others. For Foulcault (2003), it is necessary to reject a single, easily interpreted explanation, insisting on the existence of a multiplicity of meanings and senses, generally hidden - a pertinent understanding in the use of discourse analysis. With regard to sources through discourse analysis, Fischer's understanding is conveyed:

> For Michel Foucault, you have to stay (or try to stay) simply at the level of existence of words, of things said. This means that you have to work hard with your own discourse, letting it appear in the complexity that is peculiar to it. And the first task in achieving this is to try to let go of a long and effective apprenticeship that still makes us see discourses only as a set of signs, as signifiers that refer to certain contents, carrying this or that meaning, almost always hidden, disguised, distorted, intentionally misrepresented, full of real intentions, contents and representations, hidden in and by the texts, not immediately visible. It is as if within each discourse, or in a time prior to it, the truth could be found, untouched, which is then awakened by the scholar (FISCHER, 2001, p 198).

To understand discourse analysis is to understand its tools, to understand the binomial in which the variables of discourse operate - the exclusion of discourse and the ordering of discourse. This binomial can also be understood through three procedures: the limitation, rarefaction and subjection of discourses, all of which are subject to internal or external factors (FOUCAULT, 1996). In this way, the author suggests three tasks for an efficient discourse analysis: questioning the will to truth, and therefore to power; restoring discourse's character as an event; and suspending the sovereignty of the signifier, in other words, abandoning the univocity of the assigned meaning.

To this end, Richardson indicates objectivity, systematisation and inference as research techniques. The three stages are distinguished as follows: Objectivity consists of concisely and clearly defining the objectives and categories, as well as correctly defining and identifying the textual elements. Systematisation refers to the exhaustion of available sources and materials, both those that corroborate and those that deny the hypotheses raised by the researcher, in a systematic way and without exclusion by personal choice or by focusing on the interest in seeing a particular hypothesis confirmed. Finally, Inference is summarised as the acceptance of a preposition on the basis of other prepositions already accepted as true (RICAHRDSON et.al. 1999).

Bardin (1977) points to content analysis as a framework of useful and necessary knowledge for understanding communications:

> Content analysis is a **set of techniques for analysing communications**. It is not one instrument, but a range of tools; or, more strictly speaking, it will be a single instrument, but marked by a great disparity of forms and adaptable to a very wide field of application: communications (BARDIN, 1977, p.31).

The technique requires the use of hermeneutics or deductive analysis, preferably. For this study, the option of discourse hermeneutics through different perspectives is appropriate, mainly because it attempts to circumvent one of the main characteristics of speech raised by Foucault, according to which discourse presents itself as

> [...] a set of anonymous, historical rules, always determined in time and space, which defined, at a given time and for a given social, economic, geographical or linguistic area, the conditions for exercising the enunciative function (FOUCAULT **apud** FISCHER, 2001, p.204).

Hermeneutics in discourse analysis seeks the meaning of speech not in isolation, but in comparison with the socio-historical elements that complement it. Thus, it is not a question of re-signifying the discourse, but of understanding it in the light of the social and historical context in which it is inserted.

3.1. 2Discourse Analysis through Legal Hermeneutics

For the hermeneutic analysis, it was necessary to pre-analyse, select and analyse the materials and, finally, compile and process the results obtained from the discourses identified, always observing the risks to which the material is subject:

> I suppose that in every society the production of discourse is at the same time controlled, selected, organised and redistributed by a certain number of procedures whose function is to conjure up its

powers and dangers, to control its random occurrence, to avoid its heavy and fearsome materiality (FOUCAULT, 1996, p. 8).

In analysing legal norms, we sought to understand them from a legal methodological point of view and, to this end, we had to refer to the ideas of Hans Kelsen, an important legal philosopher in terms of the positive legal order. In his work The Pure Theory of Law, Kelsen (1998) seeks to understand the nature of law itself, as well as its relationship with society, morality and the other sciences.

Kelsen (1998) seeks to answer ***the*** question "what is law?", and concludes throughout his study that it is a normative, coercive and unique order, pointing to the intrinsic nature of law:

> The epistemological (theoretic-gnoseological) answer of the Pure Theory of Law is: under the conduct of presupposing the fundamental norm: we must conduct ourselves as the Constitution prescribes, that is, in harmony with the subjective meaning of the constituent act of will, in harmony with the prescriptions of the author of the Constitution. The function of this fundamental norm is to substantiate the objective validity of a positive legal order, that is, of the norms, laid down through human acts of will, of a globally effective coercive order, that is to say: to interpret the subjective meaning of these acts as their objective meaning (KELSEN, 1998, p.225).

Norms are hierarchically organised, with lesser laws The Federal Constitution, which represents the will of society, determines its guiding principles.

There is a clear separation between Law, which is absolute, and Morality, which is relative. It is not, therefore, the role of law to legitimise and interpret society and its needs, but only to describe and order it, while the other social sciences are responsible for interpreting the designs and wills of the collective (KELSEN, 1998).

Hart (1996) identified recurring questions in legal theory that deserve attention because they were not answered satisfactorily by the jus-positivists and jus-naturalists. For Hart, these were recurring questions:

> How does law differ from orders based on threats and how does it relate to these? How does legal obligation differ from moral obligation and how is it related to moral obligation? What are rules and to what extent is law a matter of rules? (HART, 1996, p.18).

Contrary to the positivist understanding of the coercive nature of the norm, for Hart (1996), the Law is divided into primary and secondary rules, in which the former determine socially accepted behaviour, based on customs and traditions, of a pre-legal nature, while the secondary rules regulate, alter and judge social conduct, establishing limits and both public and private powers. It is from the combination of these two types of rules that the foundation of the legal system itself is born, organised in terms of recognition, obedience and compliance, with the rules being accepted or not by the society on which they are imposed and regulated (HART, 1996).

Evolving Hart's concept, Dworkin (2003) points out that the maintenance of a legal system triggers an interpretative attitude towards the values invoked by this same system. In other words, the system itself emanates and stems from a set of values that imbue and are imbued by the norms they solidify, where the subject of the system must, in addition to interpreting, assign meanings and re-signify the legal system in order to give it value. The adoption of an interpretative stance is necessary and, although Hart points to the existence of obscurity in the norm in relation to the specific case where interpretation by the judges is allowed, Dworkin is categorical in repudiating such a position, and the judge is not in charge of discretionary interpretation. They must be guided by the guiding principles that underpin the legislation and the legal system in order to make a correct judgement. Therefore, there is a normative consistency which confers integrity to the system, comprising the spirit and the will that permeates and constructs a given system (DWORKIN, 2003).

A legal hermeneutic seems to meet the needs of the study, although some interpretations may not be appropriate, as they correspond to systems that are too open or subjective. Nor will a positivist legal analysis along the lines of Kelsen be useful, since it is self-contained and far from interpretative.

We therefore adopt Luhmann's (2007) understanding, which suggests that Law should be understood as a theory of social communication, transforming it into an open, extensive, autopoietic text (ROBLES, 2005). For Luhmann (2007), law is a social communication system, in fact, one of the three pillars of society, also supported by the economy and politics. In this set of systems, they are defined by the meanings that emerge from their interiors, in filtered environments within a hypercomplexity that exists within these groups, through autopoiesis.

A closed system is thus formed, given the filtering of its hyper-complexity, and at the same time it is open in the face of interactions with other existing spheres. Therefore:

> With regard to law in particular, Luhmann challenges both the analytical doctrine (positivism), because it is too fragmentary, and the hermeneutic doctrine, because it is too open; for him, law is a social system whose function is the stabilisation and congruent generalisation of normative behavioural expectations, having, like other systems, the function of reducing the complexity of being (of psychic systems) and of social behaviour (FANK, 2013, p.678).

We are thus moving towards the understanding that law is, in its essence, communication and, therefore, text. As Robles states,

> [...] there are a number of more technical or philosophical reasons that have brought semiotics (the science of signs) and hermeneutics (the science of understanding) up to date. Both branches of knowledge proclaim a broad conception of the concept of text, according to which text is not just written text, but any reality susceptible to interpretation (ROBLES, 2005, p. 26).

To instrumentalise this interpretation, there are three possible analyses: the study of the structure of language, where a logical-linguistic analysis takes place; the study of meaning through semantic analysis and; the pragmatic study, where the dynamic aspect that leads to the creation of the text is given voice. Thus, within Law, Robles (2005) points to three strands: the formal theory of Law; the dogmatic-legal theory and, finally, the legal decision theory.

The three theories pointed out by Robles refer to the three strands listed above, seen in the conceptions in which Kelsen, Hart and Luhmann stand out. The author asks whether the theory of law and the forms of interpretation would bring it closer to philosophy, also pointing out that law is a system of social communication (ROBLES, 2005).

3.1.3The Hermeneutics of Environmental Law

In terms of the hermeneutics of law linked to the environmental area, Belchior (2009) highlights the need for a democratic rule of law so that a discussion can arise about the environmental good to be protected and supported by legal studies. Thus, the need for the law to adapt to the society to which it belongs is affirmed:

> Law must adapt to society (through induction in favour of justice), just as society must conform to law (through deduction, guaranteeing legal certainty). This is the classic balance that must occur in the Democratic State of Law (BELCHIOR, 2009, p. 150).

Belchior relates the Rule of Law to principles such as solidarity, legitimacy and juridicity, listing them as major principles under which the other principles will be organised and will dictate the rules, giving them the necessary characteristics to shape the proposed society (BELCHIOR, 2009).

The principle of legitimacy stands out, supporting the interpretation of the other principles as fundamental and inherent to the rule of law, which aims to become an environmental state.

Here, the concept of the Environmental State is understood as the one coined by
Michael Kloepfer (1989 p. 5): **"A state that makes the** safety of its environment **its task, as well as the criterion and procedural goal of its decisions". In** addition,

> [...] **it evokes associations with the concept of the "social state" (environmental state** as a state that strives for a balance between social demands on nature on the one hand and the preservation of the natural bases of life on the other, and that helps to shape this balance = environmental state), which (still?) has an essentially positive connotation [...] (KLOEPFER, 1989, p. 6).

It is in the emergence of a new type of state, or a new concern and priority for it, that principles take on a structuring role, because just as they provide guidance for the legal system in force, they mould it as well

The author therefore contributes to Environmental Hermeneutics by understanding that its concepts and principles remain in constant collision and juxtaposition in the face of the clash of collective and individual wills. Therefore, the author contributes to Environmental Hermeneutics with the understanding that its concepts and principles remain in constant collision and juxtaposition in the face of the clash between the wills of the collective and the individual, sometimes in conflict, which must be disciplined by the intervention of the State. The state, in turn, must adapt to the needs of its citizens in order to correctly interpret and apply the rule (BELCHIOR, 2009).

In this sense, regarding the transformation of the state, Nunes Junior states:

> This implies the emergence of a new state and a new citizenry, who are fully aware of the planetary and indiscriminate environmental devastation caused by development, thus aspiring to new values such as ethics for life, the rational and solidarity-based use of natural resources, ecological balance and the preservation of genetic heritage (NUNES JUNIOR apud BELCHIOR, 2009, p. 165).

The conceptions of discourse analysis previously seen in Bardin and Foucault are brought together, organising them in such a way as to systematise the complex thinking seen in Morin (2007). The complexity paradigm is analysed, which is based on the concept that modern knowledge is watertight and full of errors, precisely because it is segmented, fractionated into hyper-specialised areas that don't communicate.

The formation of complex paradigms, i.e. supralogical principles for organising thought, entails a risk for hermeneutic understanding, as they result in the isolation of objects from what surrounds them, making them incapable of understanding the whole and the fractionated at the same time, or the unitas multiplex (MORIN, 2007). Abandoning a simplifying view, or even a reductionist view, is not only appropriate, but also necessary for understanding the socio-eco-organisational perspective as proposed by Morin for complex, integrated and multidisciplinary thinking. It is necessary to realise the emphasis that must be placed on interactions and to abandon the reductionist notion of the umwelt[1] , in order to adapt the vision to the complexity required, otherwise the qualities and properties inherent in the object observed in a watertight manner will not be perceived as a whole (PENA-VEGA, 2003).

The relationship between the environment and the system makes for a complex analysis, which does not allow for a disjunction between these elements. It is necessary for there to be a continuous process that refers to a living organisation, which is involved in complete and permanent vicissitude, either through disorganisation or reorganisation of these factors. Thus, it is necessary to "complexify" thinking in order to understand the relationship between the environment and today's society (MORIN, 2007).

These changes are also perceived by Leff (2003), for whom, due to the catastrophic changes that have occurred, the global ecological crisis is not the result of a natural change, but rather a transformation of nature induced by modern metaphysical, philosophical, ethical, scientific and technological conceptions: "complexity emerges as a response to this constraint of the world and nature by ideological, technological and economic unification" (LEFF, 2003, p. 19).

---

[1] **Unwelt is** understood here as the internal environment, which acts on each subject, not being an objectively determined physical or biological environment, but a subjective environment that is conceived **from a perceptual field of the "receiver of meanings"** and which, in turn, interacts with it in an operational and effecting field. (HIDELBRAND, 2013, p. 2239)

The proposal for a complex environmental epistemology prevents the adoption of Kelsen's positivist paradigm or a holistic interpretation through the hermeneutics of the whole that disregards the parts. Thus, Luhmann's view is that the Law depends on filtering to reduce its complexity, while Morin indicates that anthroposocial phenomena do not respond adequately to rules that are less complex than those required for natural phenomena themselves (FANK, 2013).

Returning, then, to the construction of the proposed environmental knowledge, it is clear that this occurs in a complex, integrated way, beyond the limiting boundaries of the traditional sciences that can extend to Law as a Science. Thus, Fank (2013) brings together, organises and systematises the thoughts of Morin (2007), Leff (2003) and Castells (2010):

> It is imperative that the approach to law (and all the social sciences) takes into account the definition and/or identification of the nature of the social force within a complex ecosystem. It is not enough for the legal-environmental paradigm to internalise technical knowledge about the environment, nor to understand its own impact on the environment; it must embrace the dynamic and generative relationship between the two. The communication between Law, Politics and Economics (the Luhmannian social subsystems), in the construction of meaning for each of them, needs to be informed by the feedback from the Environment on social life noted by Morin - it needs to serve a project of social reconstruction from the recognition of otherness (FANK, 2013, p. 682).

This provides the basis for an epistemology of environmental law that goes beyond analytical hermeneutics and comes closer to Morin's complex thinking. Once the discourse analysis of Bardin and Foucault has been adopted, understanding that law is communication and text, as proposed by Robles, makes it easier to understand the need for an environmental state, as proposed by Belchior, which points the way to environmental legal knowledge along the lines described by Leff.

We therefore used a hermeneutic approach capable of appropriating the explicit or implicit meaning of the data, in order to contrast the representations and practices aligned with those discourses.

3.1.4 Representations in Hermeneutics

Although chapter one of this work gave a preview of representations of the environment, it is important here to go into more detail about the sociological concept adopted for the hermeneutic study.

Representation, for Chartier, indicates the discourse and conceptions of certain tendencies, the appropriation of reality constructed and negotiated by these segments, thus constituting a differentiation between them.

In order to understand the process of creating public policies on environmental education in Palmas and how they internalise external elements (LEFF, 2010), it is necessary to identify the social representations that exist within official discourses, seeking to find the points of convergence between the intentions of appropriation and the practices carried out. To do this, we need to understand the representations of these groups through the socio-environmental notions they have adopted and how they re-signify, appropriate and use these representations within their discourses (REIGOTA, 2007).

One of the great challenges for the hermeneutic compression of environmental representations lies precisely in the qualification of the relationship between humans and nature. It is therefore essential to include the human being as a biological entity, without mischaracterising the cultural anthropological aspects, in order to build a new relationship with the environment and with nature (PIEPER et. al., 2012).

With regard to the hermeneutic study of Environmental Education in the light of the theory of social representations. Condenanza and Cordero point to the existence of three distinct fields of dispute, which are worth highlighting:

> **Although we can identify some institutions, organisations, intellectuals and activists, we still don't have any rigorous studies on the players in the field of environmental education at regional or international level. Therefore, we began by differentiating three subfields in which interests are disputed by different groups of actors. These interests are not determined once and for all, nor are the ways in which they are fought for: strategies can vary according to the correlation of forces, and changes in habitus are conditioned by the very structure of positions in a field. The three major sub-fields correspond to: (i) the macro-institutional, state and interstate spheres (which include the education system and its official curriculum, as well as environmental management policies); (ii) activism more closely linked to civil society; and (iii) academia and the scientific field. We could say that the field of environmental education has been built up in the *processes that have taken place "from above" (alluding to instances of the state structure, with a higher degree of institutionalisation), and "from below" (referring to the initiatives of local actors, or of organisations that play a role more closely linked to exerting pressure - through direct action, the massing of demands and/or insistence - on state bodies than to fighting them from the inside), and which has also been built up through scientific production, which has crossed both processes*** (CONDENANZA and CORDERO, 2011, p. 7-8). 7-8).

Having identified the methodology of discourse analysis and how we intend to study the disputes concerning the manifestations of representations and their practices, it is necessary to cut out and define the object under study.

3.2 The Profile of the Object - Analysing Legislation on Environmental Education (EE)

The intentionality of the Palmas legislature in relation to Environmental Education is fundamental to understanding it in relation to the basic guidelines in the light of the major legislation. To this end, it is imperative to analyse the environmental and urban planning legislation in force, especially given the proposed changes to the Municipal Master Plan, in view of the necessary adjustments, either to comply with aspects of the Municipal Executive or to comply with the Statute of Cities.

The study analysed the Municipal Master Plan, which is in the process of being renewed, depending on the completion of the legal procedures, including consultation with the population, the preparation of reports and approval by the Legislative Branch. It is in this process that the disputes between the different interests of the groups involved find an appropriate place, and where the confrontations could be appropriated by Environmental Education as an argument and the appropriation of the participants in relation to the theme.

The geographical focus of the research was the municipality of Palmas, the capital of Tocantins, a city defined in this way on the IBGE website:

> Planned to be the capital of Tocantins, Palmas was created on 20 May 1989 (shortly after the creation of the state) and installed on 1 January 1990, after the transfer of the provisional capital, Miracema.
> The first step towards planning the definitive capital was taken shortly after the election of Governor Siqueira Campos, on 15 November 1988, who requested a survey to define the location of a city that would enable it to be a hub for the state's economic and social development. The results of the study determined an area located between the municipalities of Porto Nacional and Taquaruçu do Porto, to the east of the village of Canela.
> The establishment of Palmas was only possible with the transfer of the administrative headquarters of the municipality of Taquaruçu to Palmas, making the elected mayor of Taquaruçu, Fenelon Barbosa, the first mayor of Palmas. With this decision, Taquaruçu do Porto became a district of Palmas, as did Taquaralto and Canela.
> The name Palmas was chosen in honour of the district of São João da Palma, the seat of the first separatist movement in northern Goiás, and also because of the large number of palm trees in the region (IBGE, 2016).

Situated between the right banks of the Tocantins River, now dammed to form the Palmas Lake, and at a distance of approximately 15 kilometres from the Serra do Lajeado as its eastern and western borders, Palmas is also bordered by the Água Fria stream to the north and the Taquaruçu Grande stream to the south, covering 11,085 hectares and with a capacity to house 1.2 million inhabitants (TEIXEIRA, 2009).

According to Aires,

> the phytophysiognomic characteristic is marked by cerrado grassland vegetation, cut from east to west by the Ribeirões: Taquaruçu Grande, Brejo Comprido, Sussuapara and Água Fria, whose gallery forests were defined in the Master Plan as green areas, forming large green patches in the urban layout; by the dry sub-humid tropical climate defined by dry and rainy seasons, with an average temperature ranging from 22° to 28° C; the flat to gentle terrain that ensures good horizontal land use, a fact that has conditioned the plan into a linear form, as well as the natural limits imposed by the Lajeado mountain range and the Tocantins river, which have contributed to the city's good urban and landscape setting, according to Fighera (AIRES, 2010, p. 46). 46).

It is important to note that although the architectural and urban aspects of the city are not the main focus of this study, understanding the dynamics that characterise Palmas helps us to understand the forces at work in the city. The last capital to be built has been the subject of numerous studies, most of which highlight the city's environmental and urban planning and characterise it as imaginary, artificial, proposed to be timeless, post-modern and a symbolic creation, as can be seen in Valéria da Silva's studies:

> The meaning of the city and the conjuncture that gives it specificity, that establishes an image of

> the city of Palmas, is part of the trajectory of this work, but not only that. We seek to understand it fundamentally as a city
>
> imaginary absent of time by the very context that gave rise to it. These conditions would give the city of Palmas a post-modern condition. From this point of view, the urban space is read in the wake of the theorisation of plastic time; thus, in addition to long time, short time, slow time and fast time, we discuss absent time as one of the frames that frame the projected city (SILVA, 2010, p. 32).

The ecological character of the city is emphasised in the architectural and urban design, as Aires points out:

> [...] also emphasises that the architects tried to take into account the environmental aspects of the place, seeking an interaction between urban implementation and the prevailing ecological conditions: climate, topography, biological diversity, vegetation and hydrography. In fact, in the city's Basic Plan, which was later altered into the definitive Master Plan, it can be seen that there was a concern to develop a city in harmony with nature (AIRES, 2010, p. 45).

The importance of the environmental issue for the planning, creation and development of Palmas can therefore be seen, a process strongly influenced by the historical context arising from both the 1972 Stockholm Conference and the 1987 Brudtland report. This context emphasised the socio-economic and cultural issues of a global society, which demanded solutions to problems arising from a lack of planning, as well as answers to the problems of population growth and ecological imbalance (FIGHERA, 2005).

The time frame spans two distinct periods. The first, relating to the establishment of the city, recalls the original urban planning and the development of the pre-Rio-92 master plans. It also covers the period from the promulgation of the organic laws to the effective policies and actions related to Environmental Education. The second period corresponds to the actions of recent years related to the proposal for the new Master Plan and the resumption of popular participation and political-environmental thinking.

The creation of the state of Tocantins mirrored the historical moment in which it took place. Together with the Federal Constitution of 1988, which valued fundamental rights, it influenced the consolidation of legal texts and the administrative structure of the new state. The laws and regulations took into account the provisions of the Federal Constitution, rather than having to adapt to a new reality. In the same way, they influenced the structuring of Palmas. Therefore, the process of creating the municipality is inseparable from the emancipation of the state of Tocantins, as there was in fact the creation, from scratch, of an artificial and planned metropolis, which had social, political and symbolic functions for the new federal unit.

Palmas emerged as a representative landmark of symbolic power, above all of success in the historical-political process of a country that was struggling to awaken from a period of exception. Not for nothing, its planning was permeated with mismatches and idiosyncrasies that harked back to the old caciquismo, as opposed to the original proposal to follow the determinations contained in urban planning documents such as the 1989 Athens Charter systematised by Le Corbusier (SILVA, 2010).

The strength of the symbolic discourse is evident in the following record:

> And with three days to go before handing over the government, he inaugurated the Palace and handed over the key to the other government. This act of his was very representative, because the other government may not even have understood it, when he said: I'm handing over the administrative headquarters of the State of Tocantins - this act was very symbolic, because he said: it's here and nobody else will take the capital, he meant it, because in reality, around the palace it was all shacks (SILVA, 2010, p. 59).

There was a clear concern to consolidate a political locus as much as an urban one. Although all the planning was carried out by a commission, it had to adapt to the needs of the discourse, making the practice throughout its installation period very different from what had been planned. There was a clear mismatch in the historical occupation process between the discourse enshrined in law, especially in the urban planning study, and that conducted in practice by the Executive Branch, which, among other issues, was not guided by the process of expansion and development of the urban occupation areas, which should have been guided by zoning (TEIXEIRA, 2009).

The constant regional political disputes have not given way to urban planning over the years. There has been a series of revisions and readjustments to Palmas's Master Plan and environmental policies in a constant clash between legality and the political will of the Executive's managers (TEIXEIRA, 2009).

A long road has been travelled for the title of "ecologically designed capital" to be recognised. Although the discourse would have you believe so, there was a big discrepancy between the desired title and the adjustments needed to comply with the Cities Statute. Although the desire to recognise Palmas as an ecological capital was common in the speeches made, there was a clash between groups linked to education and the environment that contested the merits of this recognition.

Therefore, diverging from previous studies in the area, which focus on the existing dichotomy between discourse and practice in Environmental Education and the perception of pedagogues from public and private education networks, such as Aires (2010), Leite (2016), Machado et al. (2011), Xavier (2007), Bispo (2005), Duarte (2012), as well as a large number of monographs produced within the Education, Pedagogy and Environmental Management courses, the present study deals with praxis and legal discourse in a more comprehensive way, less limited to aspects of the pedagogue's perception, and more focused on the issue of positivisation versus effectiveness itself.

3.3Identification and Nature of the Research

Once the object of the study and the methodological path to be followed have been outlined, the nature of the research and the methodology employed in this study are pointed out.

The research was carried out by exploring, describing and explaining the elements that made up the discourses and practices related to the object. With a qualitative approach, the direct source of the researched data was sought in the environment. Thus, the characteristics of qualitative research are that

> the researcher is the key instrument, the environment is the direct source of the data, it does not require the use of statistical techniques and methods, the data is analysed intuitively by the researcher, it is descriptive in nature, the result is not the focus of the approach, but the process and its meaning (GODOY, 1995B; LUCIA SILVA; MENEZES, 2005), in other words, the main objective is the interpretation of the phenomenon being studied (FREITAS and JABBOUR, 2010, p. 3).

In qualitative research, as explained above, the focus is on understanding the characteristics of the situations, events and organisations involved, without being concerned with measuring them through a stratified sample.

The focus of the research then becomes interpretation and, in this case, discourse analysis through hermeneutics. So, as Moreira puts it,

> the focus is on interpretation rather than qualification: generally the qualitative researcher is interested in the participants' own interpretation of the study situation: emphasis on subjectivity rather than objectivity: it is accepted that the search for objectivity is somewhat inappropriate, since the focus of interest is precisely the perspective of the participants: Flexibility in the process of conducting the research: the researcher works with complex situations, which do not allow for an exact a priori definition of the paths that the research will follow, orientation towards the process rather than the outcome; recognition of the impact of the research process on the participants.

> research on the research situation: it is recognised that the researcher influences the research situation and is also influenced by it (MOREIRA apud DUARTE, 2011, p. 70).

Initially, when systematising the work methodology, an exploratory stage was carried out, the aim of which was to determine whether or not the hypotheses listed for the questions posed to the object had occurred. In this sense, Godoy states:

> When the study is of a descriptive nature and the aim is to understand the phenomenon as a whole, in all its complexity, a qualitative analysis may be the most appropriate. Even when our concern is to understand the web of social and cultural relationships that are established within organisations, qualitative work can provide interesting and relevant data. In this sense, the choice of qualitative methodology is made after defining the problem and establishing the objectives of the research to be carried out (GODOY **apud** FREITAS e JABBOUR, 2010, p. 2).

After the exploratory phase and adjustments to the definitions of the problem and the scope of the object, the choice of documentary and bibliographic sources containing the data to be collected and analysed was determined.

Decoding the chosen documents was the key procedure for grasping the deeper meaning of the communications and intentions contained in them. We therefore looked for inferences about Environmental Education and Environmental Awareness in the selected documents, especially in the legislative regulatory documents relating to urban space and Education in the municipality of Palmas.

After the first exploratory stage, with a clearer definition of the object and the problem to be tackled, the selection of sources took place in two distinct phases. Firstly, a bibliographical review of the formation of Palmas and the state of the art regarding environmental discourse and the political and educational formation of Palmas. At this stage, it was possible to identify various academic works relating classroom work and the perception of educators and students on the subject of Environmental Education in Palmas municipal schools. There were also mentions of environmental projects developed by the municipality and various articles and papers related to the configuration of Palmas' urban space, especially in the 1990s. The second stage of source selection focused on legislative documents and projects related to the environment, education and urban space, especially those related to the object. As the data was collected, it was identified and organised into related themes, in this case referring to Environmental Education and Environmental Awareness.

After the exploratory phase and the selection of sources combined with the bibliographical review, it was once again necessary to choose the perspectives that would be directed at the object of study in order to adjust, according to the initial surveys, the lenses that would allow the study to continue. We opted for a more detailed spatio-temporal analysis of the object, through a geographical historical analysis, in order to better characterise and situate the reader in the particularities of the place of study.

With the data already collected, analysis was carried out with the aim of understanding it, confirming or denying the research questions and expanding and systematising the related knowledge (MINAYO, 1994, p.69). To this end, three categories were established for analysis: reference to Environmental Education; environmental awareness and the Environment.

The definitions of these three categories subsidised the discourse analysis in the ways explained above. The inclusion of Environment as a category was due to the search for references, within the discourses, of concern for the Environment, even if indirectly, even if not linked to Environmental Education, but which could indicate traces of concern for it, above all, within the discourse of the representations.

In order to carry out the discourse analysis, the materials collected were analysed, seeking to identify the key elements within the discourse, with a focus on Environmental Education, Environmental Awareness and the Environment. The search for the key concepts within the texts and sources chosen allowed us to critically analyse the discourses and observe the extent to which they came close to the practices perpetrated. Hermeneutic confrontation, as Minayo (1994) argues, corresponds to the proposal for qualitative interpretation of the data collected and develops contextualised dialogue, within the field of the historical and totalising specificity of speech and the meaning that discourse possesses, which in the study served to interpret representations and the ways in which this discourse extended into practices (MINAYO, 1994).

CHAPTER 4

RESULTS: ENVIRONMENTAL EDUCATION - BETWEEN DISCOURSES AND REPRESENTATIONS

4.1 The Palmas Masterplan and the Environmental Question

The first documents chosen for analysis were the Palmas Project and Master Plan. The choice was made because these documents allow for a historical analysis of the transformation of the discourse. In addition, given the recent and controversial proposals to change them, presented by the Municipal Executive, this option allows us to verify, throughout their validity, the issues that went beyond the discourse and those that only remained at the level of ideas.

The idea of erecting a city in the middle of the geographical emptiness of the cerrado, where the state of Tocantins would be formed by the will of political movements, was in line with the latent spirit immediately prior to its constitution. Little time had passed since the emergence of Goiânia and Brasília as modernist and planned cities, which sought to differentiate themselves in discourses and representations from the urban models that existed in Brazil until 1950 (MACHADO, 2011).

In response to the wishes of the political group, a central region of Tocantins was chosen for the new capital, with the justification that this way the city would not be orbiting the centres of power established in the south of Maranhão and in the mining region of Pará, and at the same time would be far away from Brasília and Goiás, as the opposite would de-characterise the intention of emancipation (TEIXEIRA, 2009).

> The option to build a new city was gradually asserted. The strategy, as well as seeking to escape the political disputes consolidated in the cities chosen as alternatives during the selection process, aimed to intervene in a positive way in the organisation of Tocantins' territorial space. Strategically, Governor Siqueira Campos identified a 90x90 km quadrilateral, located in the geographical centre of the state, as the most favourable geopolitical area for the construction of the capital. He also ordered that studies be carried out to pre-select areas so that the state's constituent assembly could choose the most suitable one. The report, which involved aspects of topography, hydrography, landscape, availability of infrastructure, among others, defined four areas with the potential to build a city the size of a state capital. Two of them were on the left bank and the other two on the right bank of the Tocantins River (TEIXEIRA, 2009, p.92).

In terms of intentionality, urban planning projects for new capital cities are linked to the time and historical ideology in force at the time of their conception and foundation. There was a search for the ideal city according to those values, in order to overcome any physical or social problems (REZENDE *apud* VELASQUES and RANGEL, 2012).

The question of the emergence of Palmas from a demographic vacuum, as the official discourse on the establishment of the capital would have you believe, is criticised in the studies of Velasques

(2011), who denounces the discourse denying the region's past as part of the symbolic power game in the dispute between representations.

This dispute aimed to give overwhelming weight to the autonomy of Tocantins and to the success of the leader of this conquest, establishing as a symbolic landmark the victory of a leader's obstinacy. The crowning achievement would be the creation of a city without the weight of past attempts and failures, without the weight of decadence and economic abandonment, only as a model of the future and modernity. A city detached from any inglorious past.

Thus, the urban space of Palmas, since its creation, has served different degrees of usefulness and commitment to the conflicting representations. The supposedly ecological character attributed to it was supposed to fulfil an environmentalist discourse, but at the same time it needed to support the discourse of political power and domination of the abandoned west.

In the construction of a discourse that would allow the development of a timeless city, an ecological perspective was adopted, which would prioritise quality of life and respect the assumptions contained in the Athens Charter. Xavier (2007) puts it this way:

> In short, the planning proposal for Palmas aimed to innovate, but in reality it failed to go beyond existing urban planning proposals, forming a combination of various currents. It was covered with the ecological discourse of environmental quality, with that of comfort and leisure for the inhabitants, with that of flexibility in the construction of the blocks, but in reality its entire construction was involved with the construction of the road system. In other words, it is really a decalogue, albeit with good intentions, of the rationalist proposals following the Corbusian, aesthetic-road and Athens Charter tendencies (XAVIER, 2007, p. 53).

Having overcome the question of the intentionality of the foundation of the new city in ideological terms, once the proposal to serve a discourse of power was clear, the question of environmental identification was also overcome, which sought support and validation in the New Athens Charter and in the pre-Rio-92 proposals for comfort and integration with the environment. It remained to discuss validation in terms of discourse, and to do so we used an analysis of the laws that first made up the urban plan, which was only later transformed into the Master Plan.

Thus, the city of Palmas was not born with a Master Plan, but with an urban plan, dated 1989, under the aegis of Law 31, which defines the Municipal Code of Works. This law, which was actually intended for Taquarussu do Porto, was soon revised and enacted by Law No. 45 of March 1990, whose purpose was the same as the previous one: to regulate municipal works. Most of the provisions of this law refer to the Master Plan and the Urban Zoning Law, both of which did not exist at the time and were only enacted in 1994 and 1993, respectively. It is noteworthy that at no point does the building code deal directly with the environment, it only makes reference to waste disposal issues in a succinct manner and without further details.

Regarding the landscape, there was a provision and concern related to the urban aesthetics of the new capital, set out in article 7 § 8, according to which the City Council could refuse to approve projects that presented deficiencies in terms of hygiene and comfort for residents, or aesthetic solutions that were inconvenient to the urban landscape (PALMAS, 1990).

Article 310 of the same law also ordered the demolition of anything that harmed urban aesthetics, but at no point did it clearly define what constituted a desirable landscape or aesthetic standards. Analysed through the aforementioned modernist discourse adopted in the founding of the city, this legal prerogative could well serve to indicate opposition to anything that offended or harmed the environment, but in an arbitrary and legally insecure way.

We found no reports of demolition actions that had condemned works in Palmas for aesthetic or landscape reasons at the time. Therefore, it can be inferred that the existence of these provisions was more to adapt the legislation to an urban environmental discourse than to an effective practice by the Public Administration.

While there was still a mention of the requirement for a green area, but only in relation to a building whose purpose was to house a hospital, the aforementioned code required that 1/10th of the area be set aside to maintain a wooded area.

The purpose of the Municipal Code of Conduct (Law No. 371 of November 1992), expressed in its preamble, stated its intent:

> The purpose of this Code is to establish rules governing public hygiene, public welfare, the location and operation of
> commercial, industrial and service-providing establishments, as well as the corresponding legal relations between the Municipal Government and the Municipalities (PALMAS, 1992).

Although issues of aesthetics, conservation, standards, distances and parameters for afforestation were delimited, there was no express concern for the environment in this code. In fact, when the new city was established, the native forest was cleared, replaced by ornamental plants or plants foreign to the Cerrado biome, such as the mango tree, a leafy, resistant, fast-growing tree that, due to its dense canopy, prevents the survival of the surrounding native forest.

There were also a series of considerations that referred to the Master Plan, which was still non-existent but promised to be regulated in the future, leaving open questions about architectural aesthetics. There were still no regulations for articles 182 and 183 of the Federal Constitution, which deal with urban policy and assign responsibilities to the Municipal Authority, which should oversee and guarantee the performance of the social functions of urban spaces as well as the well-being of its inhabitants.

Article 182 § 2 of the Federal Constitution states that "urban property fulfils its social function when it meets the fundamental requirements of **city** planning **expressed in the master plan" (BRASIL, 1988).** However, until the time of its foundation, Palmas was not required to have a Masterplan, as the municipality did not have the minimum of 20,000 inhabitants.

There were also no delimitations contained in the Statute of Cities, only a series of bills, among which Bill 4.310/1989 stands out, which proposed, albeit timidly, to provide for national and regional spatial planning plans, with an emphasis on development policy and urban expansion and the environment.

Returning to the political discourse involved in the creation of Palmas, Fighera (2005) points out that, in order to constitute the urban plan, it was planned to set aside a green area in order to promote environmental issues and fulfil the social function laid down in the Constitution:

> Within the ecological park, a botanical garden and a zoo were planned. This park would be located to the east of the city, in the Ecological Reserve between the sources of the Água Fria stream and the Brejo Comprido stream, covering an area of 10 million square metres, intended for leisure and the enhancement of existing nature. The authors considered that the best
>
> he way to guarantee the preservation of nature would be through the organised and harmonious occupation of the Ecological Park, combined with the development of ecological awareness - of human/nature coexistence. The proposed botanical garden envisaged the cultivation and research of native cerrado species and the maintenance of a plant nursery, in order to meet the demands of the new capital and the leisure of the population. It also aimed to become a centre for environmental education (FIGHERA, 2005, p. 93).

The Zoning and Urban Land Use Law (Law 386 of February 1993) sought to regulate areas of the city. It also endorsed the understanding of the contextualised discourse of the time, stemming from Rio-92, the Convention on Biological Diversity and the imminent New Athens Charter that was to be published in 1993. Thus,

> In an interview, the president of Naturatins at the time the **city was** set up **said: "This city** was born to be the ecological capital of the country. Many specialists and technicians are working towards this and it will take a lot of effort to combine the environment **and social development"** (PERFIL, 1991, p. 20). This interview highlights both international influences, which sought to reconcile preservation with economic and social development, and national influences, such as the above-mentioned discussions in the National Congress on improving quality of life (FIGHERA, 2005, p. 99).

The Zoning and Land Use Law made express provision for green areas, but only in a timid way and through a few provisions. Of these, in section IX, green areas, only two articles were able to refer to the constitution of these areas, defining them as those for the preservation or creation of vegetation cover and protection of water resources, and providing for the practice of leisure and recreational activities in them, as long as they were compatible. The law made no provision for environmental education, which was mentioned as a concern in the city's constitution. The green areas, which should also be home to and provide for integration and environmental education, were only provided for as leisure and sports areas, and nothing else was mentioned or earmarked in the legal text.

There was then a gap, a distance between the discourse of intent, which proposed the creation of the Palmas Ecological Capital brand in the year 2000, and the practical reality established through the legal system. Although there was still provision for a green area and even the creation of a council for Architecture, Urbanism and the Environment, there were few guarantees and legal provisions for the implementation of the Environmental Education discourse, which remained on the fringes of the laws and present only in the speeches that capitalised on the Rio-92 moment.

It took time for the Master Plan to be defined as an urban planning instrument, with the aim of organising the development of both the physical and urban aspects, as well as in terms of the economy and social function of a city or urban territory, promoting well-being and social justice.

Urban planning, with the aim of complying with the provisions of the Federal Constitution, has shown a concern for complying with principles that guarantee a state of democratic social welfare under the rule of law, including human dignity and a healthy and balanced environment. Therefore, in the view of public prosecutor Karin Sohne Genz:

> In short, it is an instrument through which the municipal government, acting strictly within its sphere of competence (art. 30, inc. VIII, and art. 182, § 1, both of the Federal Constitution), establishes the rules for the proper control of the use, subdivision and occupation of urban land (GENZ, 2007).

The doctrinal understanding for legal practitioners, especially those whose primary function is *custus legis,* i.e. acting as enforcers of the law, is that the Master Plan cannot be compared to a pro forma law, but rather considered a guiding instrument for urban organisational practices in line with constitutional principles, including environmental protection.

The Palmas Master Plan was instituted by Law No. 468 of January 1994, and more clearly defined issues related to green areas and environmental preservation zones. With it came definitions of important concepts for legal purposes, such as macro-parcelling and green areas, according to article 6 of the law. In this Master Plan, the term green area did not refer to natural areas, but was defined as

> [...] public or private property, delimited by the City Hall or indicated in the plans and memoranda of subdivision, destined for the implantation or preservation of afforestation and landscaping, aiming to ensure environmental and landscape conditions, and may be partially used for the implantation of Urban and Community Facilities (PALMAS, 1994).

Evidently, no care was taken to maintain the original area, but rather to opt for a landscape project full of symbolism and representations, committed to the political ideology that determined the founding of the capital.

Another relevant article concerns the allotment process, establishing that the project should be submitted to the city council for analysis of compliance with the planning guidelines, and also requiring the opinion and approval of the municipal environmental body - the Palmas Nature Foundation (EcoPalmas) - in accordance with its own regulations and Decree-Law.

The Palmas Master Plan law, enacted in 1994, has been amended four times, the last time in

2009. Both in the original body of the text referring to the urban plan and in the Master Plan, there was no consolidation of a significant environmental concern or even one that referred to the founding discourse. Thus, it is clear that the intention to make Palmas an ecological capital has not managed to transcend the plane of ideas and be recognised in legislation.

> Thus, as the city was implemented, there was a fragmentation between the ecological discourse and the practice, and it can be seen that even in the face of environmental criteria, the construction of roads and the zoning of activities prevailed. In an interview, the planner explained that the **humanistic** ecological dream **was part of "government propaganda, and** Palmas **was given** the epithet of ecological" (XAVIER, 2007, p. 85).

Even the environmental practice related to the constitution of the city was no more than a landscape plan and an effort to transform the natural environment so that it represented the political will and discourse of modernity and independence, to the detriment of the natural environment.

Environmental education, which was supposed to be a concern linked to the preservation zones, didn't go beyond the realm of discourse in the initial plans. The documents were deliberately vague, being regulated by complementary laws, decrees and later additions, thus allowing, during the initial years, total freedom for the management and aesthetic adaptation of the capital to the needs of its founders and creators.

The Municipal Organic Law of Palmas (Law No. 48 of April 1990) underpins and regulates municipal guidelines, complementing the laws previously mentioned and even establishing priorities and standards to be obeyed by the lesser laws subordinate to it, such as the Master Plan.

Article 179 of the Organic Law states that the Master Plan should be approved by the City Council and that it should be the basic document for urban development and expansion policy. Therefore, it would be from this particular instrument that all the guidelines for the composition, utilisation and planning of the urban network would emanate (PALMAS, 1990).

The same law also determined that the social function of property would only be respected if it complied with the provisions of the Master Plan and respected its cultural and environmental integrity. Article 179, § 2 also states that:

> The Master Plan, drawn up by the municipality with the participation of organisations representing the community, will cover the whole of its territory and should contain guidelines for land use and occupation, zoning, urban indices, areas of special and social interest, economic and financial guidelines, administrative guidelines, nature conservation and environmental control (PALMAS, 1990).

Nature conservation and environmental control were listed as important elements of the Masterplan, both in the law and in the founding speeches. But what was observed was a huge gulf between the discourse of representations and practice. Even with the express provision for the preservation of nature, the Palmas landscape project tried to eliminate a large part of the native forest in favour of urban development.

When the environmental discourse met with resistance from the development discourse, it's easy to imagine which of the two prevailed, even under the blessing of the public administrators of the time.

> Precious elements raised in the Urban Plan were disregarded, such as concerns for the natural environment. The landscape made up of species native to the cerrado (savannah), with a vegetation of crooked and sparse trees, was aesthetically inappropriate to be maintained in the face of a splendid new city that was being born. Thus, machines and men were put on a collision course in the face of this unpleasant physiognomy in the eyes of the new capital (FIGHERA, 2005, p.102).

In the discourse, there was confusion between the concepts of environmental preservation, respect for the environment and sustainability with the aesthetics and afforestation of the region. Furthermore, the integration plan and the ecological discourse were neglected in favour of the aesthetic construction that best suited the power project linked to the creation of the new city. The parks, preservation and integration zones envisaged in the urban planning project and in the architects' ideas never got off the ground, but were replaced **by the need for "progress" and** urban development.

> The heart of Brazil, adorned with the rich vegetation of the cerrado, beckons to itself the air of development, which should guarantee the population of its land decent living conditions, but the much-desired progress could jeopardise the ecological balance of the region (PERFIL, 1991, p. 20).

There was recognition of the failure of the initial plan to make Palmas an ecological capital, since the concern for sustainability and integration with nature had given way to urban needs and the aesthetics of the discourse of power.

> In fact, this prediction has been confirmed, as this progress has annihilated much of the native vegetation of the cerrado, entire avenues have been torn up, removing all the vegetation found. On Avenida Teotônio Segurado, the main thoroughfare linking the city in a north-south direction, spaces that are now the centre of the avenue show replanted vegetation, while others show no vegetation at all, because the native species that could have been preserved have been devastated. The degree of degradation is evident, characterising a development process that was initially based on a vision of poorly sustained progress (FIGHERA, 2005, p. 100).

The city, a huge construction site in its early days, did not meet the standards of beauty and conservation. Soon, the native forest was sacrificed, causing other environmental imbalances, such as soil erosion, especially with the removal of riparian forests. As early as 1991, Naturatins pointed to the environmental degradation resulting from disorganised growth and the failure to comply with the initial proposal, warning of the need to adopt measures to restore the environment.

> With regard to the preservation areas, the initial idea of converting them into linear parks also remained on paper. This evidence is worrying, because since 1991, the then president of Naturatins, in an interview, has warned of the devastation of a large area of vegetation, in view of

> the construction of temporary bridges in these places that were causing soil erosion (PERFIL, 1991 ***apud*** FIGHERA, 2005, p. 103).

The environmental discourse was not the only one to suffer setbacks with the establishment of the city.

There was also a discrediting of related social issues. In practice, if Palmas did not come close to Brasilia in terms of architecture, it did so in terms of social segregation. As for the occupation of urban space, financial and property pressures and interests have overturned the ecological discourse in the name of capital speculation.

Xavier (2007), in an interview with the Secretary of Urban Development and Housing of the municipality of Palmas, collected testimony about urban fragmentation by caste. Ironically, the informant refers to part of the population as migrants, but aren't they all migrants, since the city didn't exist before? We can see that the discourse of the outsider, the migrant, the invader, is still perpetuated in these lands, promoting class inequality to this day, since the city's origins.

> According to him, Palmas was planned to be between the Brejo comprido stream and the Sussuapara, but at the beginning, the government of the time went against the plan and separated the blocks by class, it didn't unite the employees [...] The least qualified employees went to Taquaralto [...] Those with the highest level stayed, the deputies and secretaries stayed in ARSE 14. Then he opened up ARSE 51, which was not part of the original plan [...] coordinators, advisors and heads went to ARSE 72 and then the lots were drawn in 1990 for the ARSES with the end of 1 [...] ARSE 61,71,81,91,101,111 then it went linearly to the south, and disregarded the original plan that foresaw the city being built in this central area, so all the problems we face today began with not following the original plan, which foresaw that 200,000 people, the entire population of Palmas today, could be between these two streams [...] the biggest problem in Palmas today is disorganised growth, disorganised growth generates all the other problems [...]. Also the people who came to Palmas without resources, they didn't allow these people to stay in the central area (XAVIER *2007,* p.88).

The city's constitution was not in line with the official discourse, nor with what was laid down in the municipal laws. Among these, I would highlight the aforementioned Organic Law of the Municipality of Palmas (Law No. 48 of April 1990) which, by providing for the Environment in its own chapter, defined in general terms, albeit vaguely, the concerns and intentions of the Executive Branch on this issue.

Article 182 of this law comes very close to the constitutional provision on the environment when it states:

> Everyone has the right to an ecologically balanced environment, an asset for the common use of the people and essential to a healthy quality of life, imposing on the municipal government and the community the duty to defend it for present and future generations (PALMAS, 1990).

The law also sets out the ways in which the municipality must act and protect the environment and, among the stipulations, is the promotion of environmental education. However, even though the discourse analysis points to key words related to caring for the environment and promoting environmental education, as can also be seen in the letter of the law, the practice is controversial, neglecting the positive text when compared to the practice and reality in the early years of the capital.

It should also be noted that the Palmas Master Plan, as well as its Organic Law, although they conflict with each other when it comes to the environment, are artificial constructions that were demanded by the pressure of a modernist discourse influenced both by the 1988 Federal Constitution and by the repercussions of the Rio-92 conference. If the tone of the discourse in that historical context had been different, this would also be reflected in these documents. However, the practice of what was intended to be consolidated with the construction of Palmas could not be denied or diverted by transitory or, rather, fashionable issues. There was an expectation related to the founding of Palmas and it was clear: to fulfil the desires and discourse of the caste that was in power in the new state.

4.2 The Cities Statute and the new Palmas Masterplan

The second moment of analysis of the Palmas Master Plan takes place in 2006, when Complementary Bill No. 14 was proposed for the new Master Plan, which gave rise to Complementary Law No. 155 of December 2007, which sets out urban policies for the municipality of Palmas.

The proposal for a bill aimed at revisiting both the Master Plan and the Municipal Organic Law stemmed from a regulation at federal level, when, more than a decade later, the constitutional provision took shape and the Cities Statute was enacted by Law No. 10,257 of July 2001.

The Statute of Cities was an important document whose goal was to stabilise the process of growth and urbanisation, encourage popular and democratic participation in this process, pay attention to the relationship with the environment and prevent the ever-present speculation in real estate.

> Finally, the Neighbourhood Impact Study (EIV) is considered a contemporary instrument because it seeks to balance the environment. It was inspired by the previous environmental impact study, but differs from it in its application. This instrument democratises local decision-making and makes public consultation possible as a requirement for urban planning and municipal building permits. Above all, it focuses on analysing essentially urban issues when reproducing space, such as: population density, the impact on urban and community facilities, traffic generation, demand for public transport and other issues that affect the artificial and natural environment (BAZOLLI, 2012, p.111).

With the enactment of the Cities Statute, the Executive had to

This was because the original Master Plan was out of date with federal legislation. Above all, there was a need to regulate this democratic participation, especially of low-income populations, who exert great pressure on the city's urban formation.

> Urban interventions on the outskirts of the city, where the low-income population is predominant,

> can lead to such an increase in property values that the municipal government may not wish to see, namely the displacement of the humble population to places that are increasingly distant from the urban centre. In these cases, it is up to the local government to be vigilant to prevent property speculation, demanding compliance with the social function of property, expressed in the law of its master plan, gradually using the measures provided for in § 4 of art. 182 of the Federal Constitution (HARADA, 2004, p. 79 ***apud*** BAZOLLI, 2012, p. 106).

However, the democratic and conscious participation of the population, especially the low-income population, could not be expected when educational instruments were not proposed. It can be seen that this group of excluded people, who live on the margins of the urban space created and valued by property speculation, exerts a great deal of pressure on the executive branch, since they demand decent housing conditions and occupation of the territory, while economic issues prevent them from doing so.

Both Xavier (2007) and Bazolli (2012) and Coriolano (2013) point to social exclusion as one of the causes of the existence of large demographic voids in the central regions of Palmas, while at the same time there has been disorderly densification in sensitive regions, damaging the environmental relationship. The labour force needed for the emergence and growth of Palmas was not suited to urban aesthetics. As a result, this working class was displaced to more remote areas, lacking in urban infrastructure and very poorly valued.

There is a direct relationship between family income and the distance of the settlement from the urban centre, indicating that the lower the population's income, the further south they were positioned, where land and settlements were more precarious and cheaper, as they were far from real estate speculation.

Only families with an income of more than four minimum wages were able to break through the social barrier imposed by the market. This social segregation has resulted in countless complaints and related studies, but they have had little effect in terms of circumventing the existing situation (CORIOLANO, 2013).

Popular participation was introduced by the Statute of Cities in Article 2, item XIII. For this reason, it was necessary to revise Palmas' Master Plan and environmental law in order to accommodate the legal provision.

> [...] hearing the municipal government and the population concerned in the processes of implementing undertakings or activities with potentially negative effects on the natural or built environment, the comfort or safety of the population (BRASIL, 2001).

But how can we expect the population to participate effectively and not just pro-forma in issues involving the development and urban occupation of the municipality? How can we expect the population to understand environmental impacts and their consequences when the historical process of occupation took place outside their participation, guaranteeing social exclusion for economic and political interests?

The reform of the Masterplan brought with it several changes in this regard. The Master Plan contained in Law No. 468 of January 1994 made no provision for the democratic and participatory practices that were to make up the text of Complementary Law No. 155. In comparison, the innovations brought in by the complementary law significantly altered the way in which the Executive and Legislative branches should promote urban occupation, as long as they respected the legal provisions.

For Lira (2011), the need to change the original masterplan was due more to the mistakes made in the original planning, which had been disfigured to such an extent that it had to be adapted:

> [...] the government's greed to profit from the sale of plots of land has contributed in a very marked way to Palmas' growth becoming one of the biggest urban planning blunders in Brazil and perhaps the world. Palmas represents the greatest example of disorganisation, of urban chaos, sponsored by a savage capitalism that never measures the consequences of its savagery (LIRA, 2011, ***apud*** BAZOLLI, 2012, p.261).

There was an effort to understand citizenship as a fundamental element in the discussion and implementation of the Master Plan, both on the part of federal legislation and on the part of the municipality. Articles 9 et seq. of the complementary law therefore provide for citizenship, sustainability, socially just, environmentally balanced and economically viable development as goals, through democratic management with the participation of the different segments of society, both in planning and in execution and monitoring (PALMAS, 2007).

The great innovation contained in Complementary Law No. 155 concerns a chapter entirely dedicated to the environment, as well as various provisions throughout the text that refer to the environmental issue, both the relationship between society and urban space and the environment, and its protection.

As well as complementing the Palmas Master Plan, this law also complements the provisions of Municipal Law No. 1,011 of 2001, which deals with Environmental Policy, Ecological Balance, Preservation and Recovery of the Environment.

The chapter deals with environmental planning, the management of environmental resources and assets, sustainability and urban landscaping.

Another innovation of this complementary law concerns the Special Zones of Social Interest (ZEIS), established by article 25. These include areas for specific and exclusive use for the construction of social housing, with urban planning parameters regulated by specific law, as well as parameters related to social inclusion (PALMAS, 2007). Likewise, the law establishes the Special Zones of Environmental and Tourist Interest, complementing the previous article in terms of not only allocating urban space for the settlement of the population, but also for the rational use and preservation of environmental space, in a complementary way.

However, the legal provision did not prevent the clash of interests between the population's desire to live close to the centre or to seek more favourable settlement conditions in the face of the real estate speculation that dominates Palmas (SOUSA, 2015).

The discourse takes up the initial idea from 1990 of promoting social justice and guaranteeing a place for everyone in the country's youngest capital, promoting integration between the urban and the natural and guaranteeing the population's access to property. However, almost 20 years after the city was founded, its urban and social geography was already well defined, with established pockets of poverty and property development, which in themselves represent mechanisms that are adequate enough to prevent the lower-income population from accessing prime areas.

In this way, the question posed earlier was only sidestepped: what is the point of calling for the participation of the city's population if they are not given the proper tools to discuss environmental and urban issues? Even though the complementary law provided for the promotion of environmental education aimed at social and urban development, little was done at the time.

Bazolli (2012) concludes that the analysis of urban sprawl in Palmas was the result of a historical process of land appropriation for purely speculative purposes, and that the municipality itself sought legal alternatives to maintain the process of valuing the areas in the wrong way, without obeying the necessary technical criteria, which caused a stir involving various segments of society that embargoed the actions and called for a participatory process.

These intended changes, draft alterations and amendments to the complementary law were eventually interrupted, since in 2016, due to Resolution 25 of 2005 of the Council of Cities, the process of renewing and restructuring the Master Plan for Cities began, which was scheduled to take place every ten years. In the work plan presented, it was determined

> effective participation of public authorities and civil society in all stages of the process, from drafting to defining decision-making mechanisms, according to Article 3 and Paragraph 1 of the aforementioned Resolution (PALMAS, 2016).

In terms of the provision of Environmental Education, the Complementary Law also determined that it should be developed in conjunction with the programmes and guidelines for socio-economic and training activities, which would be regulated by a specific law.

It is now time to discuss and analyse the educational measures adopted to bring discourse and practice closer together, through legislation and projects aimed at environmental education and raising people's environmental awareness.

4.3 Legislation and Practices in EE in Palmas - 1990s

With regard to education and environmental legislation in Palmas, there are three phases: the first in the city's constitution, the second following federal regulations after the Convention on Biological Diversity and after Rio-92, which influenced the Educational Guidelines and the National Curriculum Parameters, and the third in the re-discussion of the Master Plan, especially after 2010.

In terms of Environmental Education, it is possible to draw a parallel between the regulation of the Municipal Organic Law in the light of the Federal Constitution and the legislative changes resulting from federal laws after 1999. These changes were the basis for Law 1.011 of 04 June 2001, which laid down the municipality's Environmental Policy. Similarly, Resolution No. 2 of 15 June 2012, which establishes the National Curriculum Guidelines for Environmental Education at all levels of education, also influences Palmense's legislation, where there was previously a gap on the subject.

The Palmas Municipal Organic Law of 1990, in the chapter that deals with Education, contains a provision that determines the responsibility of the municipal entity to promote Environmental Education at all levels of education, as well as public awareness for the preservation of the Environment. This section concerns the legal determination that imposes on the Municipal Government and the community the duty to defend the ecologically balanced environment for the common use of the people, in line with the constitutional provision seen above (PALMAS, 1990).

The same law, when dealing with health care, mentions the need for health awareness as a duty of the municipality's primary schools, a provision that was later changed to health education. Although it does not address the same issue as environmental education, the proposal to promote a healthy and sustainable environment unites the two educational perspectives.

The provision contained in the Organic Law was generic and broad enough to comply with the 1988 Constitution, as well as the ecological discourse that surrounded Palmas at the time of its creation. Checking the practices carried out in those early years, it is not possible to ascertain whether, in fact, environmental education was promoted or even the ecological preservation advocated at the time of the city's foundation.

In 1991, Palmas created the Palmas Nature Foundation (Funap) through Law No. 88 of 1991. However, Funap's defined objective was not to promote environmental education, but only to protect the environment and prevent degradation. The organisation was supposed to act as a technical government body responsible for gathering data and drawing up reports on environmental issues.

In terms of state legislation, Decree 1.011 of 15 May 1990 instituted the Environmental Education Programme in the state, under the tutelage of the Tocantins Nature Institute, Naturatins, in compliance with the provisions of article 225 of the Federal Constitution, without any major innovations or differences in responsibility and with a vague proposal aimed at preserving and conserving the environment. At the time, the legislators understood that environmental education actions would not be of a formal nature, and for this reason it would not be up to the education sector to meet this demand. However, one can see a contradiction in the legislators' perception that all educational issues should also fall within the scope of the Department of Education (Seduc). This would not exclude the activities of other government sectors for this purpose, but it would confine responsibility to those who have a right and obligation to it, Seduc, which was confirmed years later with the National Policy for Environmental Education.

The law that deals with state public policies related to the environment is No. 261 of 20 February 1991. According to this law, Naturatins, then still a private law body linked to the Public Administration, was to carry out inspection, control, licensing and judgement of infractions of works and activities related to or that could pose a risk to the environment. In addition to defining Naturatins as the body responsible for environmental inspection and control, the law also defines the state's responsibility as promoting environmental education. It also determines that one of the functions of the environmental police is to act in favour of and through Environmental Education, and establishes the transfer of financial resources to Naturatins for exclusive use in activities to develop and support Environmental Education, as well as for the execution of public policies related to environmental policy in Tocantins (TOCANTINS, 1991).

In 1992 the Environmental Engineering course was created at the University of Tocantins Foundation, Unitins, in partnership with Naturatins, Ibama and the State Planning Secretariat. This partnership also led to the first formation of the State Council for Environmental Education in 1993.

Albeit timidly, actions were beginning to emerge that proposed overcoming the environmentalist discourse of political representations and working more towards everyday practices. However, what we saw was the proliferation of other discourses, with little potential to leave the world of ideas in the face of existing budget restrictions.

Naturatins was in charge of environmental education at state level, and even these were included much more as a mere formality than to be effectively implemented, with timid proposals for environmental education, since the focus of public managers, both municipal and state, was on development.

Even before the provisions relating to environmental education were mandatory, it can be seen that the legislature, foreseeing the need to prepare the municipal legal system for the imminent changes, tried to provide for such issues. It can be seen that, although they were not observed in practice, the provisions on the subject were enshrined in law.

The initial years of the capital of Tocantins revolved around the constitution and consolidation of the new city. Therefore, the acts and actions aimed at the environmental sphere were related to landscaping and urban occupation. Major infrastructure works were needed to accommodate Palmas and, at times, environmental protection was an obstacle to development, as in the case of the hydroelectric power station, bridges, roads and avenues (FIGHERA, 2005, p. 105).

Of particular note is the Luís Eduardo Magalhães Hydroelectric Power Plant (Lajeado

Hydroelectric Power Plant) project, which required the preparation of an Environmental Impact Study, resulting in a series of reports, including Volume IV, dated November 1996, which dealt specifically with environmental programmes and their conclusions.

In order to mitigate the effects of the construction of the hydroelectric dam, various environmental plans and projects were drawn up for the affected communities, including those in the municipality of Palmas, seeking not only to allocate these groups and compensate them for the material losses resulting from the flooding of the reservoir, but also to promote environmental education.

Themag Engenharia, the concessionaire responsible for the environmental impact study, indicated that one of the projects to mitigate the damage caused would be to carry out an environmental education programme, justifying that the proximity of Palmas and Porto Nacional would enhance such actions:

> The proximity of Palmas and Porto Nacional, two important urban centres, will enable the implementation of research and environmental education programmes. In addition, consideration should be given to the need for conservation measures in an area with a tendency towards growing anthropisation, both in the surroundings of these centres and along the highways, a fact that will be accentuated by the construction of the Lajeado HPP (THEMAG, 1996, p.58).

Regarding environmental education as a mitigation measure, he linked it to the development of programmes to preserve the environmental surroundings and the presence of archaeological sites as effective measures.

Based on the proposed programmes, the environmental feasibility of installing the plant was determined, since the biggest impact would be the displacement of part of the region's population. It was also considered that the creation of conservation units, management of flora and fauna sampling, and the carrying out of an archaeological survey would be sufficient to mitigate the impacts.

The developmentalist discourse ran roughshod over the environmental discourse of integrating society and nature. Even with the mitigation proposals pointing to the need for Environmental Education, considerations about the impacts were minimised until it was acceptable for the project to be approved, since its implementation was fundamental for the economic development of the region.

However, the document generated to analyse the impact of the plant records that, only six years after the foundation of Palmas, urban development planning had already been compromised, replaced by disorderly growth when the population arrived to occupy the outskirts of the cities, largely due to property speculation that prevented settlement within the planned urban centres.

The disorderly occupation in itself generated a situation of environmental imbalance in the region, aggravated by cultural, social and economic problems (THERGASA, 1996, p. 97). In this way, the proposal to promote environmental education as part of the implementation of the Lajeado Power Station was aimed at resolving both the impact of the project and the problem arising from existing irregular occupations.

The planning of the Environmental Education proposal took into account federal legislation, in this case the current Forestry Code set out in Law 4.771 of 1965, Law 5.197 of 1967, which provided for the protection of fauna, and Law 6.938 of 1981 on the National Environmental Policy, as well as Article 225 of the Federal Constitution of 1988.

These laws do not deal specifically with Environmental Education, but they do have similar elements, such as the obligation to include texts of forestry interest in radio and television programmes, as well as encouraging the education of the community to participate effectively in the defence of the environment. All of these actions, however, are not formal in nature, and their effectiveness is, to say the least, questionable.

At the state level, Decree No. 1,011 of 1990 provided for formal education to promote awareness through the public education network, in order to train and achieve the proposed objectives of preserving and respecting the environment. The plan included, among the subjects to be addressed, the specification and recognition of the natural and cultural characteristics of the region by the populations affected, in order to integrate society and public bodies in their responsibility for the conservation and preservation of the environment, in line with the understanding of the authors visited earlier about the concept of Environmental Education.

In perspective, the plan proposed in terms of education and raising awareness among the population revived the discourse of integrating society and nature present in the process of founding Tocantins and Palmas, not only through the proposals to create Conservation Units, but through the educational proposal and the involvement of society as a whole, in its different groups and responsibilities. In this context, the report determined:

> Partnerships with formal educational institutions and **NGOs that work with non-formal education, as well as countless other** entities, are of fundamental importance, since it is in these institutions and organisations that the multipliers of the information and actions of this programme are to be found (THEMAG, 1996, p. 61).

We can see the construction of a discourse that refers to the importance and necessity of integrating society and the environment, through projects aimed at raising awareness among the population about environmental impacts and integrated living, dealing with issues such as deforestation, fire prevention, preservation of riparian forests, among other actions that were carried out, but with little long-term effectiveness.

In terms of Environmental Education, no project or partnership has lasted longer than the legal requirement of six years. Although the plant's management sporadically promotes programmes related to the subject, both internally and externally, there is no sense of permanence, evolution or continuity in the proposals, which remain watertight and isolated, pro-forma meeting the needs of maintaining an environmentally responsible discourse (MOURA, 2011).

The municipality of Palmas included in its 1994 multiannual plan, by means of Law No. 502/92, the provision for participatory Environmental Education, which proposed that it should prioritise the stimulation of the educational process, the preservationist consistency, reaching all levels of Palmas society. The target audience for this proposal was teachers and environmental agents, and the action strategy was to train them to integrate environmental education into the curricula of the municipal education system (PALMAS, 1994).

The Multiannual Plan had a report that subsidised the proposed decisions and, among them, on social impacts, there was the deconstruction of the political discourses that indicated Palmas as an ecological and planned capital. The report mentions Palmas' housing policy and relates the problem of homelessness to the environment, but states that the municipality lacks a permanent and effective environmental policy.

> As far as the environment is concerned, Palmas had no permanent or effective environmental policy. There was no concern for the environment, either on the part of government activities, state and municipal, or the private sector. Given the need, the Municipal Executive will adopt measures aimed at protecting ecosystems. Allied to these factors, the growing disorganised migration to the state capital has contributed significantly to environmental production (PALMAS, 1994).

The discourse contained in the report shows that there was at least partial alternation in the group that held power in its different spheres. There was a need, therefore, to point out the previous administration's lack of care for the environment, as well as a commitment to effective and lasting action.

The alternation of groups in power has influenced the discourses. These may even remain similar, as in the case of maintaining Palmas as an ecological capital, but the discrepancies between discourse and practice are evident, in this case, issues related to the Environment and Environmental Education.

There is a direct relationship between environmental imbalance and the need for planning, for environmental education of the population, and a great discrepancy between the developmentalist political discourse and the founding of Palmas with the reality that was observed in its constitution.

The disparity was not limited to the issue of housing; there were also problems related to rural activities and the exploitation of the region's natural resources, which occurred in a disorderly manner, de-characterising the flora and fauna, as well as successive fires, as the report states (PALMAS, 1994).

The same Fundação Natureza de Palmas (Palmas Nature Foundation), formerly known as Funap, now known as EcoPalmas, was seen as an agent of change, working tirelessly to reverse the damage caused by means of inspection and pollution control measures. However, apart from the change of name, there were no permanent changes or plans, which the report indicated was the municipality's intention.

What was observed was the perpetuation of a discourse of **demagogic** environmental concern, such **as the use of expressions like** "EcoPalmas' **tireless efforts to** fight **for the fulfilment of its duty". But** little progress has been made in terms of legislation, apart from some insipid actions, more aimed at landscaping than at environmental education, such as setting up a forest garden and distributing tree seedlings.

One couldn't expect different legislation when the main law that dealt with education, the Law of Guidelines and Bases (LDB), No. 9.394 of 1996, made little or no reference to environmental education. The LDB treats environmental education in a vague and imprecise way, totally at odds with the environmentalist discourses that emerged from the Rio-92 environmental movements and the CBD.

> The new Guidelines and Bases Law (LDB), instituted by Law 9.394 of 30/1112/96, did not establish any provisions on EE and does not even expressly mention it. Only with a lot of goodwill can we attribute to the legislator any intention to deal with this issue, even if only indirectly. With regard to primary education, the LDB establishes that curricula must include knowledge of the physical and natural world and the social and political reality, especially of Brazil (Art. 23, § 1). Among other aims of higher education is to stimulate knowledge of the present world, in particular national and regional knowledge (Art. 43, VI). That's all and in reality it's nothing, and it could be said that on the environmental issue, the LDB has not listened to the immense national and international effort that since the Stockholm Conference in 1972 has sought to include environmental education as an instrument of public policy relevant to the promotion of a new, fairer world order (BARBIERI, 2002, p.10).

The proposal for environmental education was flawed, even on the part of the Ministry of Education and legislators, who had not yet realised the relevance and influence of the environmental movements of the time on the issue of integrating society and nature. The incompatibility of environmental discourse with legally established pedagogical practice on a national level led to a reform three years later, in order to correct the gap left in federal legislation.

Returning to the state level, the proposal for the multiannual plan for the following four-year period, governed by Law No. 681 of 10 November 1997, determined the same purposes, with very similar wording, innovating only by allocating Environmental Education as a strategy for controlling environmental pollution. It also broadened the objective to include integration with the environment, as well as preservationism (PALMAS, 1997).

In the part of the discourse that was analysed, the invocation of Palmas as the Ecological Capital was used loosely in an attempt to justify heritage care:

> Palmas is known as the Ecological Capital, due to the large number of green areas destined for environmental preservation in the city, as well as the large number of wild animals that inhabit these areas (PALMAS, 1997).

The excerpt above, taken from the multi-annual plan, was used to justify the actions of the

Metropolitan Guard of Palmas, which leads to two considerations: the need to justify and perpetuate a configuration through a discourse and the maintenance of the term Environment **as an "umbrella" capable of sheltering the most diverse interests of** the dominant political representation.

Law No. 721 of March 1998 authorised the Municipal Department of Works, Urbanism and the Environment to grant a permit and regularise works completed up to December 1997. In practical terms, it granted amnesty to works in precarious conditions, with a mere favourable opinion. Although the need for regularisation is understandable, the practice legitimised by this law undermines the concerns voiced in previous speeches and zoning plans, and the precautionary principle.

Of the actions mentioned that were not very effective, we would highlight the basic sanitation project for low-income communities, in 1995, and the diagnostic evaluation of the Environmental Education Programme in Porto Nacional and Palmas, in 1997. In addition to these, there have been some measures derived from Unitins' actions in favour of the environment, such as the Environmental Education Extension Project for the conservation of the Taquaruçu District's seaside resorts, which took place between 1995 and 2000, among various other production, research and scientific initiation projects, especially those linked to the Environmental Engineering course.

In 1995, the Tocantins State Council for the Environment (Coema-TO) was formed, as provided for in Law No. 261 of 1991 and amended by Tocantins State Law No. 791 of 1995, which went on to define the Council's functions. The minutes of its foundation, available on the website of the state environment secretariat, reveal the uncertainties of those who came to make up Coema, to the point where it was necessary, at the second ordinary meeting, to include in the minutes the need to clarify to the members what legal provision had been made by the 1988 Federal Constitution, as well as what the Council's function would be.

There was already a legal provision for Coema, as well as a legal provision for its activities. Even so, its members, who defended the discourse of environmental protection, knew little about what could actually be done through the body, denoting not a lack of interest, but a total disconnect between discourse and practice.

It was necessary to educate managers environmentally so that they realised the importance of the environmental education process as a strategy for what was proposed at the Coema-TO meetings. It was a slow process and, as can be seen in the minutes, there were actions to educate the population, but never with a precise focus, only as a complement to other plans.

The clashes with the Tocantins state government were also on the agenda of some of the meetings, as the following extract illustrates, regarding the discussion of the possible environmental impacts of the construction of the Serra da Mesa power plant.

> The Governor replied that his government is attentive to these [environmental] issues and that if

> necessary, measures will be taken, but such attitudes should be precautionary, as the repercussions of thoughtless actions could be nationwide (TOCANTINS, 1996).

Even in the face of the impacts mentioned and highlighted by the Council's participants, then governor Siqueira Campos emphasised his rejection of any measures or news that could be alarmist and have negative repercussions, which would jeopardise the implementation of the Serra da Mesa Hydroelectric Power Plant. Therefore, the developmentalist discourse continued to restrict and limit environmental practices in Tocantins.

In the long term, these actions proved to be watertight, lacking in continuity, especially those linked to public administration, which served more to legitimise a discourse on the part of the government than to focus on education or permanent action. This hypothesis is borne out by the fact that, as the parties in power alternated, so did the plans, nomenclature and focus of action, always with the idea of changing the discourse in order to appropriate the practice, **later** claiming **"paternity" of it.**

The main concern related to the environment in Palmas in the 1990s was not environmental education, but rather the analysis, inspection and release of environmental licences. Actions were also spread out, involving different institutions such as the Municipal Guard and the Federal Police, in programmes to raise awareness of preservation and tree planting. However, they were always watertight and not continuous (MOURA, 2011).

It was at the end of the 1990s that there were significant changes in the ways of working and in the positivisation of expectations relating mainly to the environment and environmental education.

The legislative changes at federal level that took place at this time, including the Education Guidelines and Bases Law, greatly affected Environmental Education and also led to changes in the legal system of Palmas and Tocantins, in a cascade effect in the legislature, given the need to adapt to the Major Law.

4.4 Legislation and Practices in EE in Palmas - the 2000s

With the enactment, at federal level, of Law No. 9.795 of April 1999, the Federal Government instituted its national policy for Environmental Education. In practice, it was determined that this would involve formal education in its sphere of action, in addition to the bodies and entities that make up the National Environment System (Sisnama). Thus, the practice of environmental education became compulsory in public and private educational institutions, in the public bodies of the Union, the States, the Federal District and the Municipalities, and in Non-Governmental Organisations working in environmental education (BRASIL, 1999).

In addition to Rio-92 and the Convention on Biological Diversity, the law was based on the Brasilia Declaration for Environmental Education, approved during the first National Conference on

Environmental Education, which adopted the recommendations of the Belgrade Tibilizi Charter as part of Agenda 21 (BERBIERI, 2002).

In line with the law's recommendations, Environmental Education also came to be understood by educational legislators as an instrument for promoting the sustainable development of the environment. The Brazilian education system had yet to be adapted to a reality that would accommodate the understanding proposed by Agenda 21, since it had a strictly technical nature.

It was not the scope of education or of the Brazilian legislator to sacrifice developmentalism in favour of a sustainable environment, as Barbieri adds:

> According to this Declaration, the existence of different concepts of sustainable development resulting from different visions on the part of different segments of society is one of the problems for Environmental Education. In addition to these, there is the development model adopted in Brazil, which favours economic aspects, the failure of the different levels of government to comply with the Agenda 21 recommendations and the lack of coordination between government and civil society actions (BARBIERI, 2002, p.10).

However, it was necessary to bring the country into line with global discourse. Brazil could not fail to provide in its legislation and practice for participation in the fight for a sustainable system, even if in practice the economic and development system adopted by the government was at odds with that proposed and expressed in official speeches. As a result, efforts were made to adapt the education law and issue decrees and resolutions that would correct the distortions between discourse and practice and still fulfil the representations.

Because of these needs, the new law had a cascading effect, requiring changes to both state and municipal legislation. The law expressly stated that the different spheres were responsible for defining guidelines, norms and criteria for Environmental Education, within their competences and in the areas of their jurisdiction, while respecting the principles and objectives of the National Environmental Education Policy (BRASIL, 1999).

It was necessary to incorporate the new guiding principles of Environmental Education into current legislation. Key principles such as totality, interdependence, pluralism, ethics, articulation, respect, multiculturalism and citizenship came to be considered crucial to the effectiveness of an environmental educational process, and should be incorporated by the teaching staff and those involved in these processes (MOURA, 2014, p. 434).

These principles, set out in Article 4 of Law 9.795/99, guide educational practices, with the aim of developing an integrated understanding of the environment and the complex relationships between the different levels of society and nature, as well as guaranteeing the democratisation of information and encouraging sustainable development, as set out in Article 5 of the same law (BRASIL, 1999).

The municipality of Palmas, through Law No. 1011 of 4 July 2001, established its environmental policy, with a chapter dedicated to environmental education. This was an improvement on the previous provision, but not a total innovation in itself. Among the changes, the municipal government, through the Environment and Tourism Agency (Amatur), was given the role of encouraging social participation in environmental issues, as well as ensuring the successful implementation of the instruments described in the law (PALMAS, 2001).

The same law assigned the municipality of Palmas the task of developing Environmental Education actions among the population, especially subsidising the actions of those who could act as multipliers of awareness for sustainability, the ultimate goal proposed for the programmes.

According to Moura (2011), in Palmas there were no projects aimed at raising environmental awareness or even emphasising the importance of preserving the environment. There were no integrated and continuous actions on the part of the Municipal Department of Education or the Environment, aimed at fulfilling the responsibilities laid down by law, apart from sparse afforestation and green space maintenance projects.

One of the proposals that came in the wake of the adaptation was the creation of the Friends of the Environment (AMA) project by Palmas City Hall. Regulated by Law No. 1023 of 5 July 2001, it stipulated that the Municipal Department of Education and the Municipal Agency for the Environment and Tourism should subsidise actions in schools for teenagers from low-income families, involving them in the mission of caring for the landscape and environmental conditions of Palmas, through training and a socio-educational dimension. The law did not hide two of its concerns: the first related to caring for landscaping, which is clearly present in the discourse and sometimes confused or used as a synonym for the environment. The second was the welfare nature of the programme, which linked the young person's educational development to receiving a grant (PALMAS, 2001).

The project was short-lived, only during the administration of that specific group, and was soon abandoned due to the costs for the municipality, as well as its low effectiveness in relation to its purpose, ending up becoming mere assistance to low-income families, similar to the "Pioneiros Mirins" programme in the state of Tocantins. Illustrating the programme's inefficiency, in an interview with the T1 portal, the mayor at the time, Odir Rocha, **said that "the AMA project** - Friends of the Environment - was another action that took thousands of children off the streets. They received a grant and looked after the **gardens. [...]" (ROCHA, 2014).**

In other words, the project that set out to integrate young people around an environmental awareness for sustainable practices ended up being reduced to financial aid and garden care. However, one cannot help but realise that the project was an effective action, and not merely a speculative one, in terms of legislation. Even if it wasn't long-lasting and veered towards assistance, the programme provided for and even carried out environmental education actions more effectively than its predecessors. It was not just a matter of visiting green areas or organising tree-planting afternoons. Behind these activities was a training proposal, but first it was necessary to "structure the structure", which lacked understanding and targeted pedagogical training.

The government tried to integrate environmental education into environmental actions by passing Law No. 1165 of December 2002 on selective rubbish collection, but this was only regulated by Decree No. 227 of 2011, almost a decade after its publication. This law stipulated selective collection as part of the Environmental Education programme to be developed in schools, with the aim of awakening ecological awareness.

Regarding the lack of selective rubbish collection, as well as the consequences of the lack of environmental awareness on the subject, Moura (2011) states:

> According to one of the experts, the population needs to reduce the amount of rubbish they produce and the town hall needs to urgently set up a selective collection system, in which a large part of the population should take part. It's clear that environmental problems have been worrying all the city's inhabitants and that's why the population should be motivated to take part in selective collection, while the town hall should treat the rubbish to avoid health and environmental problems. It would also be necessary for the city council, through the Environmental Education department, to implement a well-developed Environmental Education programme to educate the local community to change their habits by reusing waste through recycling. An additional major concern is hospital waste, which is taken to the Palmas landfill site and should be incinerated (MOURA, 2011, p.94).

It is clear that the aim of selective waste collection was to develop environmental education in Palmas public schools, and that one of the working teams provided for in the law was to specifically plan and develop environmental education actions involving collection. However, as explained above, the law published in 2002 was only regulated in 2011, and even in 2017, the city of Palmas does not have a waste collection service or centre that can be considered effective.

In 2003, State Law No. 1374 defined the state's Environmental Education policy and the State Environmental Education Information System, establishing the State Environment Council as the managing body responsible for implementing the provisions of the legislation.

The law defined the participation of the different segments of society as vital to the strategy of implementing Environmental Education at the different levels of activity. In addition, it legally defined the possibility of raising financial resources from both domestic and foreign sources for this purpose. The multiannual plan, as well as the Budget Law and Annual Budget, should contain provisions to subsidise the state's actions to implement the established environmental education policies.

A budget was then set aside to put environmental education actions into practice, which had to prioritise the development and training of human resources so that they could carry out the planning.

The law also defined the competences of the proposed policy, with public authorities being responsible for inserting new environmental policies; educational institutions for actually promoting environmental education actions, and the media for publicising related practices (TOCANTINS, 2003). Companies and trade associations were given the duty to promote actions and programmes aimed at conserving the environment, while it was up to society as a whole to develop values, attitudes and skills for individual and collective action aimed at preserving the environment; state councils to provide advice; other bodies to develop policies, and NGOs to propose and carry out programmes and projects aimed at the environment, stimulating the critical formation of society (TOCANTINS, 2003).

Municipalities were delegated the task of establishing guidelines for environmental education plans and developing programmes and actions in formal and informal education. The State Environmental Education Information System was responsible for organising, collecting, storing and processing data on the programmes developed, as well as disseminating and subsidising the actions approved for this purpose.

In this way, the 2003 law was a milestone in Tocantins' environmental legislation, defining the role of each person in the responsibilities for developing environmental education.

> The State's Environmental Education Policy is regulated by Law No. 1.374/03, which aims to conserve the environment by strengthening social values, knowledge and skills, encouraging understanding of environmental problems, indicating alternatives and the appropriate use of potential. Article 5 specifies that it is the responsibility of public authorities to include environmental education policies in all levels and forms of education, involving society in the conservation, recovery and improvement of the environment (HALUM, 2009, p. 42).

Its main merit was to treat the issue of Environmental Education as a matter of interest and responsibility for society as a whole, and not just as a watertight government policy that established criteria and lines of action to be taken. The intention was to involve the different social segments in a joint and significant action to promote and defend the environment, even if economic issues were still confined to the sphere of public power.

However, Palmas did not create a specific law for environmental education, which for a long time was linked to the Department of Social and Urban Development, and later to the Department of the Environment, although it is constantly mentioned in the legislation, bordering on issues related to formal education.

> One of the departments that the municipal environment department should have in full operation is Environmental Education, and in this sense Law No. 1.374/03 strengthens this activity and the objectives of this research. In Palmas, the public authorities have not included environmental education policies in education, so this is yet another social activity that this department should implement (MOURA, 2011, p 43).

Environmental education was seen as a social issue that should be part of the projects of different departments, which is justified if you take into account the interdisciplinary nature of the subject. It thus permeated various projects, such as carbon sequestration, landscaping, selective rubbish collection, natural preservation and even social assistance on the part of the municipal government.

It wasn't the municipality's focus to promote the environment through environmental education, but rather through various actions linked to different approaches, as in the case of the carbon sequestration project developed by the Municipal Department of Environment and Tourism in 2005. To this end, various acts were carried out for preservation, based mainly on federal legislation.

> According to the Municipal Department of Environment and Tourism (2005), the carbon sequestration project aims to carry out reforestation, preservation and environmental education actions over 20 years, by preserving non-degraded areas, reforesting degraded areas and setting up squares and gardens. Thus, the justification for this project lies in a way of preventing the degradation of the riparian forest - **Permanent** Protection Areas **(PPAs) that has been occurring intensively. Therefore, during** implementation, the project aims to transform the green areas surrounding the city's streams into Municipal Conservation Units, in accordance with the SNUC - National System of Conservation Units, in accordance with Federal Law 9.985/00 (FIGHERA, 2005, p. 174).

The creation of the Cesamar Park, by means of Law 1406 of 16 December 2005, meets the expectations of preservationism, determining the area of the Brejo Comprido water source as a biological reserve, a cultural landscaping project and a leisure area for society. Although there was provision in the city's initial urban project for the creation of parks around the streams that cut through the capital, it was only 15 years later that this measure was put into effect in the form of a law, and even then without any mention of the promotion of environmental education beyond an authorisation to do so. Although the law is still in force, the preservation aspects contained in it are ignored by the public authorities, such as the ban on using the site for events and festivities, since this would jeopardise the integrity of the biodiversity that was sought to be preserved. Even today, the site serves as a stage for events such as Christmas, parades, concerts, various activities promoted by the S System, among others, which permeate the municipal calendar and disregard a series of provisions in the law for the preservation of the environment around the park.

State legislation has also evolved with the regulation of the State Council for the Environment through Law No. 1,789 of 2007 and the State Law on Climate Change, Environmental Conservation and Sustainable Development in Tocantins, Law No. 1917 of 2008. Both brought with them regulations and changes to the legislation pertaining to the care and management of the environment that ended up influencing municipal legislation.

However, even with the legislative changes, actions within the municipal government were geared towards environmental conservation, but not towards raising public awareness. This even led civil servants involved in environmental actions to recognise that the strong point of urban environmental management in Palmas was analysis, licensing and inspection, as well as support for projects outside the public administration.

On the other hand, the same civil servants pointed out that the City Council's development of actions in favour of the environment was weak. They also pointed out problems such as the small number of civil servants assigned to deal with environmental issues and the lack of initiative and programmes related to education and environmental awareness in society (MOURA, 2011).

Moura (2011), who worked on the issue of environmental management in the municipality of Palmas in the 2000s, pointed to the dissatisfaction of neighbourhood leaders with municipal environmental programmes. In particular, she pointed out the lack of programmes aimed at the periphery and the absence of popular participation in decision-making and the development of these programmes. In severely criticising the municipal management structure, he analyses that until 2011 there was inefficiency on the part of the public authorities in dealing with the issue:

> The experts, on the other hand, said that it was essential to restructure the city council's urban environmental management, where the current model does not fulfil the aim of preserving the environment, let alone improving the community's quality of life (MOURA, 2011, p. 96).

Actions aimed at preserving the environment take place without the participation of the population, or rather, without any concern for developing a critical sense or even a feeling of duty towards environmental issues. For the most part, they have all been geared towards preserving the image and landscape of the urban space, which has served to fulfil discourses of prosperity and economic development, such as the motto of Brazil's fastest-growing and most developed capital.

Urban green areas suffer from the drought, because in most of them the landscape design doesn't take the climate into account, using exotic ornamental plants that can't withstand the drought. An example that still occurs in 2016 is the sundial garden at the intersection of blocks 304 and 404 South, maintained by the city council, where the entire vegetation is changed every three months, since only the aesthetic issue was considered.

For efficient environmental management, it should be noted that studies related to the biome should be applied, as well as the adoption of environmental actions and projects that take into account the information in an articulated manner and also contribute to the development of Environmental Education in the municipality.

At the end of 2010, there were no environmental education projects being developed by the Palmas Municipal Department of the Environment in primary schools. Only two community projects were underway, one aimed at reforesting native Cerrado seedlings and the other at recovering and protecting the springs and green areas on the banks of the Sussuapara stream.

The community-orientated project was carried out in partnership with the Ministry of the Environment and was called Palmas para o Verde (Palms for the Green). However, the programme did not include environmental education and no action was taken to raise awareness among the population. Moura (2011) also points to the ineffectiveness of the project when he observes the loss of effort due to burning by the population of Palmas itself.

> Trees were planted in the same place in January 2010, but the fires in the region destroyed most of them. To date, no environmental education project has been set up to raise public awareness of

the problems of fires. The manager of environmental education and projects said that most urban fires are caused by people burning rubbish and even arson (MOURA 82).

There were no programmes to raise awareness among the population about controlling fires, protecting water resources and managing natural resources, which are the objects of environmental education. In this sense, there were many fires and the destruction of green and riparian areas within the urban fabric, which still suffered from incorrect waste disposal.

Even today, solid waste disposal continues to be a problem for the municipality, which, despite agreeing to comply with the National Solid Waste Policy (Law No. 12305/2010), has to deal with incorrect disposal by the population, a clear example of the lack of environmental education that could be mitigated with effective awareness programmes. It can be seen, therefore, that creating laws and agreeing terms of behaviour do not solve the problems if the population remains ignorant of their responsibilities with regard to the production, treatment and correct disposal of their waste.

4.5 Environmental Practices and Discourses in the Municipal Education Plan

The project for the new Master Plan, which is still in the process of being drawn up by the municipal executive and legislature and which began in 2007, has brought a series of changes to the panorama of environmental discourse and practices in Palmas. In order to comply with the recommendations of the Statute of Cities, it was necessary to design a new proposal that was in line with federal regulations and, at the same time, plausible enough to accommodate the changes - among them, the active participation of the population in the process of drawing up the document.

In the original Masterplan, it was difficult to promote popular participation, since consultations were not common practice in the legislative process, even though they were provided for in the Constitution and various laws. So a project was drawn up in three stages: the drafting of a technical report, the report of the popular consultations and the consolidation of a Bill, in that order.

The technical project was drawn up by a non-governmental organisation hired by the Municipal Urban Development Department for this purpose. Together with the Federal University of Tocantins, the NGO Associação para a Recuperação e Conservação do Meio Ambiente (Arca) developed a specialisation course in Urban and Environmental Planning in order to train the technicians and civil servants who would carry out the reports and popular consultations to be carried out, as well as providing consultancy to carry out this work (OLIVEIRA, et. al., 2015).

As a result of the popular consultation process, the Community Reading Report was drawn up, which summarised the discussions held by the population. The result obtained through LC 155/2007 and the content of the report are not contradictory, but sometimes the latter seems intentionally vague, which ended up removing the legitimacy of the process in terms of popular participation. Even so, the democratic process of LC 155/2007 has helped to highlight environmental issues from the community's perspective.

> The community analysis carried out in the different parts of the municipality of Palmas showed, through the environmental representation, that the main environmental problems are linked to the way in which the territory of Palmas is occupied, which is contrary to what was proposed in the original project. This reading also highlighted the population's desire for the Participatory Masterplan to solve the various environmental problems in the different areas of the municipality (OLIVEIRA, et. al. 2015, p. 443).

However, the vision of society found in the document is utilitarian and pragmatic, and considers that the environment should serve society, not integrate with it in harmony. In order to reach a different understanding, it would be necessary to develop an environmental sense through awareness-raising and education, with the aim of changing existing representations, both on the part of the public authorities and the community.

LC 155/2007 attacked the consequences of the lack of socially and environmentally responsible urban planning. However, little or nothing was done to change perceptions or even tackle the causes of the environmental problems raised. There was no discussion of land use policies, as these would have hurt speculative property interests. There was no ongoing environmental education project, other than those that were legally obligatory and applied only for as long as was necessary to fulfil the legal requirement.

But what changed between 2005 and 2015? A decade has passed and a new process of popular consultation for the revision of the Master Plan has begun, and once again a clash between the practices and representations of Environmental Education has taken place in Palmas. But with what differences from those that took place in the past?

It can be seen that there was no specific or permanent focus on the development of environmental education. There were, however, legislative changes that sought to meet the demands arising from federal legislation, as was the case with the revision of the Master Plan.

In 2012, municipal law No. 1,911 of 10 August was passed, establishing the permanent education forum and determining, as its main responsibility, the drafting of the Municipal Education Plan (PME) through consultation with various segments of society, as well as its implementation, monitoring and evaluation (PALMAS, 2012).

The law established the permanent composition of a commission that would discuss issues related to Environmental Education, a theme that is present in the reports of the three meetings that took place. The first report, drawn up in 2012, uses part of the reference text on environmental education made available by the Ministry of Education and Culture, accessed in August 2012.

Among the programmes that justify compliance with educational legislation, the report cited the adoption of federal programmes related to the environment and, as a local initiative by SEMASP and SEMED, the implementation of selective waste collection in Early Childhood Education units and the Environmental Agenda Programme in Public Administration. It also indicated the implementation of COM-VIDA in all education units in Palmas, but not its role and responsibilities (PALMAS, 2012).

It was not a priority to develop Environmental Education effectively in 2012. However, in the same year, the Environmental Education competition was instituted by law 1,913 of 25 September 2012, encouraging the production of drawings, poetry, essays, dissertations and scientific articles on the subject. The aim of the prize was to raise awareness among the population of the need to adapt physical space, dispose of waste, and save energy and water, developing sustainability in practice in public schools.

With its own budget allocation and already provided for by law, there have been three editions of the competition, with prizes awarded in the categories provided for in specific notices, but it was not clear how the intended changes would take place beyond the objective of arousing interest in the subject in a cross-cutting way by encouraging it to be tackled within education units.

Once again, these were disjointed, one-off actions that were unsuccessful in what they set out to do: change the population's perception and raise awareness through environmental education. It can be seen that the actions, as a rule, lack this continuity and effectiveness, and even if on the level of ideas they express the intention of going beyond the commonplace of conventional education, they remain there.

> ...[there is] a lack of availability of specific teaching materials to address the environmental dimension, especially in the municipal education network, and (b) that the specific training for practice in Environmental Education recommended by the National Environmental Education Policy - PNEA (1999) has reached a minimal portion of the teaching staff of the school units. In this way, the environmental education practised in the schools investigated is still conventional, punctual and discontinuous because it does not reach the entire school community and does not involve local society (AIRES, 2010).

Unaware of effective educational actions, the government determined loose competences and targets, without coordinated actions that justified or indicated how it intended to achieve these results. In this sense, the Multiannual Plan for Environmental Education set the following physical targets:

Environmental Action Targets - Palmas Multi-Year Plan

Type	Action	2010	2011	2012	2013
Project	Selective waste collection units	0	0	4 und	8 und
Activity	Prevention and control of fires	0	0	100%	100%
Activity	Carrying out Environmental Education	0	0	100%	100%

Table 1
Source: PALMAS, law no. 1946 of 31 December 2012.

The approved Multiannual Plan set a target of 100% effectiveness in raising awareness through Environmental Education, but it did not specify which universe this percentage referred to or even how this awareness would be measured, which makes any statement in favour of or against the stipulated target doubtful due to the great subjectivity it contains.

As part of the permanent education forum, three municipal education conferences were held, the first in 2012 and the last in 2015, resulting in the drafting of the proposal that subsidised the Palmas PME, which was only promulgated in December 2015, almost six months after the deadline set by federal law and defined by the MEC. The discussion of the plan dragged on mainly due to issues such as gender ideology and other issues related to environmental education.

In Palmas, the vote on the law that defined the PME was still truncated, with a large number of amendments proposed by the unprepared legislator who insists on not

understand that the state is secular and therefore should not include topics that identify this or that religious book as something acceptable.

Therefore, the target involving the PME, which highlights Environmental Education, also deals with all the topics required by federal law, such as Afro-Brazilian and indigenous culture and human rights, with only the constitutional and federal provision remaining as the guiding precept for the educational practices mentioned, with the prohibition, however, of the inclusion of gender ideology and eroticisation - as the legislator considered it important to highlight.

Even though the State Education Plan, Law No. 1,859 of 2007, sets the goal of cross-cutting environmental education through integrated, continuous and permanent practices, the municipal law was largely silent on the matter. Perhaps in the shadow of this legislation, as well as the guidelines contained in the 2010 and 2014 CONAE, both of which had reference documents clearly stating the importance of environmental education for the PME.

Not by chance, at the meetings that defined the Palmas PME, according to minute 8 of the commission for the organisation and systematisation of the Municipal Education Plan of 19 June 2012, when recording the speech of technician Regina Freire Arnaldo do Nascimento, a member of Semed's Environmental Education Centre, about the absence of the topic of environmental education in the group's planning, opinions were expressed about the daily lives of the members, about the relevance and difficulty of the topic, about the possibility of debating it again in the future, and nothing more. The topic was never brought up again at future meetings, and was left as a footnote, even though the Environmental Education department had warned about it during the development of the programme.

> He then gave the floor to Regina Freire Arnaldo do Nascimento, from Semed's Environmental Education Centre, who thanked him for the invitation and said that unfortunately the PNE, the Substitute and the PEE only mention Environmental Education twice. He then explained the guiding document, Law No. 9.795, of 27 April 1999 - which provides for Environmental Education, establishes the National Environmental Education Policy and makes other provisions. She also explained what the 1995/96 National Curriculum Parameters (PCN) defined. She gave a historical overview, emphasised Agenda 21 and the concepts of the environment, sustainability,

> diversity and attitude. The importance of working from an (inter)transdisciplinary perspective [...] (SEMED, 2012).

There was a clear need to integrate and regulate environmental education within the scope of the PME, but there was no will in the discourse of representations. In fact, the environmentalist discourse lost its "charm" to the power discourse of the ecological capital: Palmas

was now the fastest-growing capital in Brazil, and its appeal was no longer focused on ecology and sustainability, but on development and economic growth.

Even so, when designing the curriculum for the primary schools grades, the technical teams considered the subject of environmental education. In the early years, it is very present, spread out among the subjects, from the first to the fifth year, mainly in Science and Geography.

On the other hand, the curriculum for the final grades of secondary school does not cover the subject continuously. Only in the subject of History, when it deals with the environmental changes resulting from the Industrial Revolution, and in the subject of Science, when environmental problems are addressed, and only in these topics, only in one module or bimester. There is no transdisciplinary or interdisciplinary conversation about environmental education. It can be seen, therefore, that the subject is approached in isolation and detached from the daily lives of students and teachers.

So, in the Palmas PME, Environmental Education is present, in a timid way, as a complement aimed at developing sustainability in a disciplinary way, with the provision of maintaining selective rubbish collection, even though this is not very effective although it has been planned and regulated for a long time.

In fact, although the PME was sanctioned in 2015, the single annex that actually regulates the matter was only published in January 2016, and as soon as it came to light, it already underwent political changes. The discussion around Environmental Education and its relationship with political discourse only became important again when the Master Plan was revised.

4.6 The Revision of the Master Plan - Discourse and Representations of EA

After almost a decade of Bill 155/2007, it was necessary to revisit the Palmas Master Plan, due to the resolution published by the Ministry of Cities. Resolutions 25/2005 and 83/2009 set a deadline of up to ten years for revising masterplans in order to bring them into line with the Cities Statute.

For the revision of the Master Plan, a work plan has been drawn up which, similar to the one carried out in 2005, includes three basic stages: technical reading, community reading and drafting the Bill. Currently, the process has finalised consultations with the community and is moving on to the preparation of the municipal diagnosis, which should reflect the wishes of the community and the technical issues raised.

These stages aim to fulfil the legal requirement for broad popular participation in public management projects. In an official statement on the Master Plan, available on the Palmas City Hall website, Impup's president, Ephin Shluger, reports that 23 meetings were held with society, defining them as fundamental for community reading:

> Through this interaction we can respond and fulfil the wishes of the population". [...] The diagnosis will give us the guidelines for the public policies that must be implemented, combined with the instruments of urban mobility and the occupation of land use, so that we have a Palmas that is pleasant to live in ten years from now," said the president (PALMAS, 2016).

As far as the methodology is concerned, we're interested in the first step proposed for community reading, which was to offer workshops to sensitise the community:

> Stage 01: Community Awareness and Levelling Workshop - consists of the way in which the processes of participation and agreeing on a reading of an existing situation should be constructed, especially with regard to the municipality's inequalities and potential. At this stage, it is important for the population to clearly understand what the masterplan is, having clarity about it as an instrument for resolving conflicts and establishing guidelines for building a sustainable city. These tools include: research, participatory observation, the provision of information to analyse reality, group techniques, stimulating the exercise of leadership and the practice of participatory planning, and training for the conquest of new spaces for debate. A Public Hearing will therefore be held to start the process of revising the masterplan, with the aim of empowering the population in the process (PALMAS, 2016).

On this occasion, there was a concern to provide the population with the minimum competences about what was to be discussed at the meetings. Awareness-raising took place along thematic lines that guided the discussions: the environment and climate change; territorial development; taxation and governance.

The methodology of the meetings included three stages: an awareness-raising workshop, when the reasons for the changes to the Master Plan and the competences of each of the axes were presented; the initial plenary session, when workshops were held to hear the wishes and considerations of the population and, finally, a new plenary session, when the population's perceptions were compared with the technical consultation, and parallels were established for the preparation of the municipal diagnosis.

In smaller communities, with less participation, the meetings were divided into just two stages, the sensitisation plenary and the concatenation of ideas plenary, thus eliminating the thematic rooms for debate, consideration and reflection on each axis.

From the meetings held, documents and minutes were compiled and are available for consultation on the website maintained by Palmas City Hall to publicise the project to revise the 2016 Palmas Master Plan.

There was a concern on the part of the Public Administration to ensure that the acts were widely publicised after the 2016 elections for the positions of mayor and councillors, when the reports resulting from these acts were made available. But this only happened after the elections, so that the results of the consultations had no significant impact on the electoral process. Once again, we see that practices end up bowing to the intentions of existing discourses and representations.

Meeting Minutes - 2ª Phase - Reading the City

N°	Document type	Location/Sector	Date Realised
01	Community Minutes	Buritirana	04/08/2016
02	Sector Minutes	Federal Savings Bank	09/08/2016
03	Sector Minutes	Fieto, Siduscon, CREA, CAU	10/08/2016
04	Community Minutes	Good Hope	11/08/2016
05	Community Minutes	Taquaruçu Grande	16/08/2016
06	Sector Minutes	OAB, CREA, CAU	17/08/2016
07	Community Minutes	St John's	18/08/2016
08	Sector Minutes	Fecomercio, ACIPA, CDL	07/10/2016
09	Community Minutes	Taquari Rural	14/10/2016
10	Sector Minutes	Roads	17/10/2016
11	Community Minutes	Urban Taquari	18/10/2016
12	Sector Minutes	CEF, Secovi, CRECI, Sindimóveis	20/10/2016
13	Community Minutes	Jaú Settlement	25/10/2016
14	Sector Minutes	Gyms	31/10/2016
15	Sector Minutes	Social Movements	03/11/2016
16	Community Minutes	ARNES	09/11/2016
17	Community Minutes	ARNOS	10/11/2016
18	Community Minutes	ARSES part I	10/11/2016
19	Community Minutes	Urban Taquaruçu	16/11/2016
20	Community Minutes	ARSES part II	17/11/2016
21	Community Minutes	ARSOS	22/11/2016
22	Community Minutes	Lago Sul, Aurenys	24/11/2016
23	Community Minutes	Taquaralto	29/11/2016

Table 2
Source: research data

The strategy of explaining, by axis, the reason for working on the Master Plan, and highlighting the Environment among them, meant that suggestions and issues relating to environmental education came to light objectively and not just subjectively, as in the consultations held in 2005. Thus, several references were made to the development of Environmental Education as a solution to the environmental problems faced by the communities.

The table below shows the survey, carried out through discourse analysis, of inferences about Environmental Education, both directly and indirectly.

Direct and Indirect References to Environmental Education

N°	Location/Sector	Direct Quote	Indirect Quote
01	Buritirana	02	06
02	Federal Savings Bank	-	02
03	Fieto, Siduscon, CREA, CAU	01	04
04	Good Hope	03	01
05	Taquaruçu Grande	02	10
06	OAB, CREA, CAU	-	-

07	St John's	-	01
08	Fecomercio, ACIPA, CDL	-	01
09	Taquari Rural	05	10
10	Roads	01	06
11	Urban Taquari	05	05
12	CEF, Secovi, CRECI, Sindimóveis	-	-
13	Jaú Settlement	03	03
14	Gyms	04	05
15	Social Movements	-	03
16	ARNES	01	02
17	ARNOS	04	07
18	ARSES part I	-	-
19	Urban Taquaruçu	08	25
20	ARSES part II	02	01
21	ARSOS	02	07
22	Lago Sul, Aurenys	03	05
23	Taquaralto	07	04
	Total	53	108

Table 3
Source: research data

To analyse the direct form, we took into account the number of times the terms "Environmental Education" **or "Environmental Awareness" were recorded** in the documents drawn up from the community meetings, i.e. recorded in the minutes.

Indirect references take into account issues relating to the environment, such as monitoring, promoting or preserving it. On average, for every three environmental issues raised, one was related to or pointed to Environmental Education as a desire of the community. Environmental Education was not only mentioned as a desire or as a solution. There were also references to the lack of it, which caused problems for the community that had no opportunities to develop this aspect.

> They also point out the lack of environmental education to provide guidance on the areas that need to be conserved and that they don't know about, reporting that they know of cases of deforestation right up to the banks of rivers and streams, and complain that they are fined for minor offences such as deforestation; they also stress the need for landowners to be correctly registered so that they can be properly fined (PALMAS, 2016).

The community's greatest aspirations can be divided into five main groups: environmental monitoring; collecting and disposing of rubbish and solid waste; conserving and protecting water resources; combating and preventing wildfires; and; emerging as a panacea for all environmental ills, environmental awareness and education, generally as an alternative to the fines and sanctions imposed by public authorities.

The participants in these community meetings expressed their desire for more effective action on the part of the municipality when it comes to environmental education. In addition, they criticised the lack of enforcement and punitive measures when there is no prior work to raise awareness among the population, considering the actions of the public authorities to be ineffective, apparently more concerned with punishing than actually preserving or solving the environmental problems that exist in the communities.

It can also be seen that they admit that a large part of the existing demands, especially those related to burning and rubbish disposal, are due to the lack of clarification, awareness and education of the residents of the various regions of Palmas. In their own defence, they claim not to have received any kind of guidance or instruction from schools or the public administration.

> 4 - I would like more attention to be paid to environmental education, rubbish collection, people dedicated to educating rural producers to plant, attention to be paid to rural producers (Weronica Raquel Soares apud PALMAS, 2016).

It wasn't difficult to find statements like the one above, with a clear interest in inclusion and improving the community through environmental education actions, which could, at low cost, reverse negligence and disregard for the environment.

When we consider the intentionality of the community meetings, we realise that the records and considerations that emerged in the communities were not free and spontaneous processes. On the contrary, they were dialogues co-opted and directed according to the wishes of the mediators, technicians and professionals representing the public authorities.

The intentionality of what is said and what is not said becomes more evident when you realise, in the reports drawn up by the Master Plan work team, that the topics debated were formatted, guided and directed in such a way as to fill out and adapt to a usable form, compatible with the technical report. A similar process took place in the 2005 consultation, but this time it happened more smoothly, more discreetly, while still in the phase of collecting data from the community.

It's not an open letter from the community to the municipality, nor is it a meeting free from surveillance or the presence of state agents. Nor are the citizens who take part unprepared. Thus, public meetings and assemblies were privileged places for disputes between representatives who held or sought power, especially if we take into account that these meetings took place in a municipal election year.

The intentions were there, and it is no coincidence that the results were not published until after the elections. It is also no coincidence that there were formal, well-crafted speeches with a cause and consequence link, attributing responsibility to this or that sector of the municipality, and denouncing the lack of ongoing projects and programmes in relation to Environmental Education.

As a result of the PME, as well as the actions envisaged by the revision of the Master Plan, changes were made to the Multiannual Plan in December 2015, with the inclusion of various actions and targets related to Environmental Education. This measure can be interpreted as a return to environmental discourse on the part of political representatives, in fact a return to discourse and proposals for practices aimed at meeting the renewed needs and demands that the image of sustainability requires.

4.7 The Sustainable Palmas Programme and the revival of an environmental discourse

In 2012, with a new change of municipal power, the current political discourse also changed. The mayor who took over the city revived the development discourse and, in need of resources, sought a partnership with the IDB, which made the intended lines of funding conditional on the development of an urban sustainability programme, in other words, the inclusion of Palmas in the list of the Emerging Sustainable Cities Initiative (Ices).

A management and planning team was soon mobilised, based at the Palmas Planning Institute (Ipup), which began to draw up the projects needed to develop the programme called Sustainable Palmas, including the revision of the Master Plan - one of Ices' requirements.

In the introduction to the programme launched in 2015, in the words of the mayor, there is a reference to the developmentalist discourse, an old acquaintance of regional practices. In this sense, the programme's introductory text highlights:

> With this in mind, the city is being planned to capitalise on its potential and vocation for sustainable growth in the various dimensions that enable economic, technological and environmental development, as well as reducing urban and social inequalities (POLIS, 2015, p. 7).

There was a revival of the discourse on sustainability and development, as well as a vocation for integrating society and nature. In order to receive financial support, a programme had to be drawn up, and for this purpose the Polis Institute was hired to coordinate the process that resulted in the published programme.

The programme consisted of presenting the challenges and goals demanded by the Ices methodology, in the form of priority areas for development, as well as the actions needed to solve their problems, all with a view to sustainability. To begin the work, a diagnostic study was carried out and three other basic studies were carried out: climate change mitigation; environmental vulnerability and risks; and urban growth. All these studies subsidised the technical and community reading of the revision of the Master Plan.

The programme presents an urban panorama of Palmas, the reasons why it has the potential for sustainable development, how it came to be classified as an emerging city and why it should be included in the ICES proposal.

> With regard to the themes in the Environmental and Climate Change Dimension, the yellow traffic light was predominant, especially with regard to air quality, climate change mitigation,

> noise and vulnerability to natural disasters, demonstrating that the environmental theme is generally not critical in Palmas. In addition, ICES has provided the municipality with studies on climate change, which will be presented in this Plan, enabling it to better position itself on the issue and provide concrete guidance on the matter. The same happened with the education indicators, which are part of the urban sustainability dimension (PÓLIS, 2015, p. 59).

The adoption of an environmental and sustainable bias in the development discourse was mandatory when signing the partnership with the IDB, and a necessary condition for joining Ices. In preparing the municipal technical diagnosis, the environmental and climate axis was a requirement, as was the stance of proposing actions and public policies capable of mitigating situations of vulnerability.

It became a condition of the partnership that a developmentalist, environmentally responsible and sustainable stance be sustained, at least in the discourse emanating from the government. Therefore, in addition to the measures contained in the presentation and justification of the project, there was also a return to environmental discourse, in many ways very similar to that adopted in the 1990s.

In addition to the speeches, there was also a return to environmental practices, some of which were educational, in order to meet the demands of Caixa Econômica Federal and the IDB in particular. These include the PME, the revision of the Multiannual Plan, projects by the Department of the Environment, the BRT tender and, in particular, the revision of the Master Plan.

Palmas, however, could not adopt environmental and/or environmental educational practices without exposing the fragility and negligence of previous years in this area. The years of predatory property speculation were viscerally exposed. In the same way, social segregation, educational neglect in schools, the lack of lasting projects and all the other wounds that have been masked over the years by the different public administrations had to be brought to light so that the necessary measures could be taken.

In terms of environmental education, it was clear that the projects to raise awareness among the population, especially about rubbish collection, a major banner of the environmental education movement, were only insipient practices, affecting only part of the population, this part being the one with the best geographical location within the municipality (POLIS, 2015).

Regarding the environmental education projects advocated over the years, the report drawn up for the Sustainable Palmas programme pointed to their inefficiency or low effectiveness. The lack of continuity was indicated as a major cause of inefficiency, especially those related to preventing fires, selective rubbish collection and preserving the environment.

In short, the environmental education activities over the years have been limited to occasional talks and planting tree seedlings, sometimes not even from the Cerrado biome. The report on the Sustainable Palmas programme has therefore exposed the discrepancy between environmental and educational discourse and practice.

Property speculation and the socio-spatial segregation of urban planning were laid bare by the report and pointed out as one of the existing environmental deficits. As a measure, the need to build a BRT transport system and revise the Master Plan was indicated. While the former gave rise to new attacks by property speculators, the latter brought the population into the dialogue, highlighting a series of new problems, including a lack of participatory public management.

> The survey of the population revealed a deficiency in the process of public participation, with 63 per cent saying they had never heard of participatory budgeting. Of the 35.5 per cent who had heard of it, 83.3 per cent had never taken part in a participatory budget (PALMAS, 2015, p.87).

As a result, in order to consolidate the intended partnership and investment, the public authorities tried to put into practice the various recommendations made by the programme, sometimes running over legal issues, which further highlighted the gap between environmental discourse and practice - see the recent case of bids and expropriations for the construction of the BRT.

As for projects and programmes related to environmental education, the June Environment Day was proposed and carried out, a month dedicated to educational activities in the city's schools and communities. The project, however, **only** brought "more of the same" in terms of proposals: the importance of selective collection - although this is non-existent or inefficient in many parts of the city - the discussion of risks and ways of preventing fires and the ever-popular planting of tree saplings.

The Environment Foundation, a body set up by Law No. 2202 of 31 December 2014, has as one of its objectives the promotion of environmental education in order to form values, attitudes and skills that enable individual and collective action in an integrated manner with programmes and projects aimed at conserving, restoring and improving the environment. It was therefore to be expected that there would be greater interaction between this body and the Department of Education, but what we saw in practice was the same form of action with isolated, uncoordinated actions and no continuity.

In this sense, the Sustainable Palmas programme has also set itself the goal of making changes to the teaching plan in order to adopt an integrated form of action between municipal bodies, thus avoiding fragmentation, lack of continuity and inefficiency in the proposed projects and programmes.

As a result, various lectures and training courses for civil servants were given throughout 2015 and 2016, as a way of meeting the goal of the Multiannual Plan and the recommendations of the sustainable development programme. However, the ecological village project in Sussuapara Park, promised since 2014, has not yet been completed, and the site remains without the renovations and adaptations needed to implement the programme.

What can be seen at this stage of transformation in Palmas' political discourse and practices are actions to meet the demands of a funding programme within a developmentalist proposal. The discourse of environmental responsibility is being revived, alluding to the need to develop environmental education, with practices that are still timid and carried out not so much out of environmental awareness, but out of a need to meet the counterpart demanded by the development agencies.

CHAPTER 5

CONCLUSIONS

The historical process of establishing Palmas as the Ecological Capital has a discourse full of symbologies and appropriations that serve different interests and political representations. The aim was to understand the forces at work within these disputes and the intentions of the discourses aimed at sustainable development in the light of the discourses and practices of Environmental Education.

Firstly, theoretical understandings of the environment, nature, sustainability, environmental representations, environmental education, urban space and environmental law were conceptualised. Based on these concepts, an investigation was carried out into the relationship between official discourse, especially that which has been transformed into law, and the practices developed by bodies and organisations linked to public administration, with an emphasis on the municipal sphere.

The first point relates to the establishment of Palmas, which emerged as a proposal to break with the dominant forces of the time, as a symbol of a new policy for a region that was allegedly relegated to an administrative and economic backwater. To justify the expense of setting up a new city, a whole project of modernity and sustainability was drawn up, demanding and justifying the emergence of a futuristic and conscious capital in the middle of the Cerrado.

Still in relation to the first point, it was noted that the environmentalist discourse was used as part of the justification, taking the hook that influenced the social, environmental and humanist movements of the 1970s and 1980s. However, the very way in which the company responsible for urban planning was contracted already showed that this was an empty environmentalist discourse, in which concerns about the environment were replaced by an idea of landscaping more suited to the wishes of those representations.

Environmental education was no different. Even with the constitutional provision, at this first moment in the capital, practices related to environmental education were relegated to those that were suitable for landscaping and transforming the chosen territory into a modern city capable of accommodating, within the segregationist policies, the immigrants who came to populate the region.

In the second point, with the development and consolidation of the city, there was recognition of environmentally mistaken practices that served economic and political interests to the detriment of the environmental agendas proposed and defended in the speeches. It was recognised, both by educators and legislators, as well as by the political agents themselves, that the practices that took place greatly disrespected the discourses emanating from these representations. However, even with this acknowledgement, little was actually done to correct the existing distortions.

Still on the subject of the city's consolidation and economic development, once the initial years had passed, environmental disregard for paving and urban infrastructure gave way to property agents and speculators who promoted real social segregation. This led to the exclusion and alienation of the **now "immigrants" from other regions, who, depending on** their means, were not welcome in the urban centre of the new capital.

However, the historical and social context was also evolving: the occurrence of Eco-92 and the documentary and legislative advances related to education and urbanism made political agents revisit the practices carried out in Palmas. Social exclusion was no longer acceptable in the way it had been, and the environmental discourse that served only to satisfy ideology was also harshly criticised.

In the third point, the evolution of environmental and educational practices showed that, despite good proposals, they were isolated, discontinuous and insipient. Some institutions and bodies responsible for developing environmental activities were set up, and the goal was to include environmental education in schools in the municipal network. However, little or no investment in material, training and effective action has meant that actions have been limited to preventing fires, landscaping, planting trees and selective rubbish collection.

As a result of the low efficiency of the preservation and environmental education programmes, some of which were more welfare than educational, the programmes were discontinued and abandoned. The population lacked access to information, as well as an efficient and effective interdisciplinary programme in schools.

It was also pointed out that the process of revising the Master Plan through participatory management, which culminated in Bill 155/2007, showed that the population was unprepared to deal with environmental issues in a conscious and coherent manner. It was noted that citizens were aware of the existence of problems within the community, but were unable to see that their long-term solution was the effective development of Environmental Education actions capable of promoting the necessary awareness. On the other hand, the official discourse claimed the existence of efficient actions related to environmental education, as well as the fulfilment of targets to raise awareness among the population.

The environmental education programmes and projects suffered from the same problem as at the beginning of the formation of Palmas: only part of the population, those occupying the more affluent areas or rural communities, received special attention due to their status and federal funding. Urban social segregation also influenced the measures taken by the public authorities, which, only occasionally and in order to meet the demands of these excluded populations, promoted small actions such as talks and planting tree seedlings in communities further away from the centre.

In the last stage of the study, we saw a return to a development discourse committed to the environment. However, we saw that this resumption is due to a demand from development agencies to participate in economic development programmes, and not by the will of the public administration or the population.

The proposal for the Sustainable Palmas programme and the city's inclusion among those taking part in the Emerging and Sustainable Cities Initiative demanded an effort from the public authorities, which culminated in the drafting and enactment of the Municipal Education Plan, which includes environmental education, as well as the ten-year review of the Master Plan, in line with the rules of the Statute of Cities.

In this last point, it was noted that the city's need to adapt to the Ices targets led to a review of some attitudes towards the environment and environmental education promoted by the municipality. As a result, there are now changes in behaviour, training programmes for civil servants, the population and public school students, as well as other environmentally conscious and sustainable landscaping and urban planning projects.

We can therefore conclude that there is a change underway, a rapprochement between the discourse and practices of Environmental Education beyond the walls of municipal schools. However, it's too early to say whether these changes will be permanent or just palliative and temporary measures to meet the needs of political agents and their representations in the face of the demand for funding and participation in Ices. A strategic analysis of environmental education, focusing on formal education and taking into account not only specific actions, but cross-cutting and interdisciplinary ones, is still far from the municipality's reality.

Putting environmental education into practice is a major challenge not only for the capital Palmas, but for the entire state of Tocantins. As such, this study does not exhaust the views and discussions about the practices and discourses related to public authorities and Environmental Education, but rather presents a political, social and historical panorama that makes further research possible. In addition, the results presented here can provide a basis for the changes proposed for the coming years on the subject, both at municipal and state level.

REFERENCES

ACSELRAD, H. **Conflitos ambientais no Brasil**. Rio de Janeiro: Relume Dumará. 2004.

ADAMS, B. G. **What is Environmental Education.** Apoema Project, published on 5th June 2005. Available at http://www.apoema.com.br/defmicoes.htm accessed on 10/07/2015.

AIRES, B. F. C. **Educação Ambiental e Meio Ambiente no Contexto de Palmas Capital Ecológica**, PhD thesis. Goiânia, UFG, 2010.

ALMEIDA, T. A. **Umwelt and Environment Modelling in Intelligent Systems.** UFSCAR, 2010.

ANTUNES, P. B. **Direito Ambiental**. 7th edition, revised, updated and expanded. Rio de Janeiro: Lúmen Juris, 2004.

ARENDT, H. **The Human Condition**. 10. ed. Rio de Janeiro: Forense Universitária, 2001.

BARBIERI, J.C. **Desenvolvimento e Meio Ambiente. Petrópolis**, RJ: Ed. Vozes Ltda, 1997.

_____________ . EE and environmental problems in **EALEGAL**, MEC, 2002, available at http://portal.mec.gov.br/secad/arquivos/pdf/educacaoambiental/ealegal.pdf accessed on 01/07/2016.

BARDIN, L. **Content analysis**. Lisbon: Edições 70, 1977.

BAZOLLI, João Aparecido. **Urban sprawl and management tools: dilemmas for local authorities and society in Palmas/TO**. Federal University of Uberlândia, Minas Gerais: 2012 available at: http://www.researchgate.net/publication/235446164_Disperso_Urbana_e_Instrumentos _de_Gesto_Dilemas_do_poder_local_e_da_sociedade_em_Palmas-TO?ev=prf_pub Accessed on .10/05/2016.

BELCHIOR, G. P. N. **Hermeneutics and the Environment**: a proposal for environmental legal hermeneutics for the realisation of the Environmental Rule of Law. Dissertation (Masters in Law). Postgraduate Law Course, Federal University of Ceará, Fortaleza, 2009.

BENEVOLO, L. **As Origens da Urbanística Moderna**. Lisbon: Editorial Presença, 1994.

BOSA, C. R.; ANDERLAN, J. **Historical approach to a forest destined for environmental conservation in the municipality of Dourados-MS.** In **Revista Eletrônica do PPGEAamb-CCR/UFSM.** Vol2, n° 2, p. 228-240, 2011.

BOURDIEU, P. **The Economy of Symbolic Exchange.** Rio de Janeiro: Ed. Difel: 1998.

_____________. **Symbolic Power.** Rio de Janeiro, Ed. Bertrand, 2002.

_____________. P. **The social uses of science**. São Paulo, Ed. UNESP, 2004.

BRAZIL. **Federal Constitution of Brazil** promulgated on 5 October 1988 / collective work authored by Editora Saraiva with the collaboration of Antonio Luiz de Toledo Pinto, Marcia Cristina Vaz dos Santos Windt and Lívia Céspdes. São Paulo: Saraiva, 2004.

. **National Environmental Education Policy** Law No. 9.795 of 27 April 1999. Brasília, MEC, 1999.

_____________. **Regulates articles 182 and 183 of the Federal Constitution, establishes general guidelines for urban policy and makes other provisions.** Law 10.257, of 10 July 2001, Brasília, 2001.

_____________. **City Statute,** Guide for implementation by municipalities and citizens. 2. ed. Brasília, 2002.

. **National Solid Waste Policy**. Law No. 12.350, of 2 August 2010, Brasília, 2010.

_____________. Ministry of Education and Culture. Resolution No. 02/2012. **National Curriculum Guidelines for Environmental Education**. Brasília: MEC, 2012.

_____________. **National Education Conference 2014** - The PNE in the Articulation of the National Education System: Popular Participation, Federative Cooperation and Collaboration Regime. Brasilia, 2014.

BRUGER, P. **Environmental education or training?** Santa Catarina Island, Ed. Letras Contemporâneas, 1996.

BRUSEK, F.J. O Problema do Desenvolvimento Sustentável In **Desenvolvimento e Natureza**: Estudos para uma sociedade sustentável. INPSO/FUNDAJ, Institute of Social Research, Joaquim Nabuco Foundation, Ministry of Education, Federal Government, Recife, Brazil. 1994.

CAPRA, F. **O ponto de mutação**. 22. ed. São Paulo: Cultrix, 2001.

. **The Hidden Connections** - Science for a New Sustainable Life. 2ª ed., São Paulo: Editora Cultrix, 2002.

CARVALHO, I. C. M. **Educação Ambiental - a formação do Sujeito Ecológico.** São Paulo: Ed. Cortez: 2011.

CASSIRER, E. **Essay on Man**. Martins Fontes: São Paulo, 1994.

CASTELLS, M. **The Urban Question.** Rio de Janeiro: Paz e Terra: 2009.

_____________. **The Power of Identity.** São Paulo: Paz e Terra, 2010.

CHARTIER, R.. **Cultural History** - Between practices and representations. Lisbon/Rio de Janeiro: Difel/Bertrand Brasil, 1990.

_____________. **A beira da falésia:** a história entre as incertezas e a inquietude. Porto Alegre, Editora Universidade UFRGS, 2002.

CLAPP, J. DAUVERGNE, P. **Path to a Green World:** The Political Economy of The Global Environment. Cambridge Mass.: MIT Press (2011) 2005.

CONDENANZA, L.; CORDERO, S. La Educación Ambiental desde la teoría de los campos In **Anais do congresso VIII Encontro Nacional de Pesquisa em Educação em Ciências (ENPEC).** State University of Campinas 2011 available at http://www.nutes.ufrj .br/abrapec/viiienpec/resumos/R1724-1.pdf accessed on 15/11/2016.

CORIOLANO, G. P. Palmas: between planning and exclusion In **Vitruvius** available at http://www.vitruvius.com.br/revistas/read/minhacidade/11.122/3587 accessed on: 11/06/2016.

CORREA, L. R. **O Espaço Urbano:** Notas teórico-metodológicas IN GEOSUL, n° 15, Ano VIII, 1 semestre, 1993.

. **O espaço urbano** - São Paulo: Ed. Atica: 1989.

COSGROVE, D. E. Towards a radical cultural geography: problems of theory. In: CORRÊA, R. L.; ROSENDAHL, Z. (Org.) **Introduction to cultural geography**. Rio de Janeiro: Bertrand Brasil, 2003

COUTO, R. LIMA, M.; KARINE, A., NEGREIROS, L. **The Environmental Education practices of the environmental management course students at the Catholic College of Tocantins Palmas/TO.** Palmas: Católica Palmas: 2010, available at http://www.catolica-to.edu.br/portal/portal/downloads/docs gestaoambiental/projetos2010-1/1- período/As praticas de educacao ambiental.pdf accessed on 05/08/2015.

CORNETTA, A. **The financialisation of climate** - a geographical approach to the carbon market and its scales of operation. São Paulo, Annablume, Fapesp, 2012.

DAMATTA, R. **A casa & a rua:** espaço, cidadania, mulher e morte no Brasil. 5. ed. Rio de Janeiro: Rocco, 1997.

DARNTON, R. **O grande massacre dos gatos**. 2 ed., Rio de Janeiro: Forense- Universitária, 1987.

DRUMMOND, J. A. A História Ambiental: temas, fontes e linhas de pesquisa In **Estudos Históricos,** Rio de Janeiro, vol. 4, n. 8, 1991, p. 177-197 - available at http://www.nuredam.com.br/files/divulgacao/artigos/Hist%F3ria%20Ambiental%20Dru mond.pdf accessed on 12/04/2016.

DUARTE, J. G. **Discursive Analysis of Environmental Education Practices in Elementary School: Case Study in a Municipal School in Palmas - TO**.
Dissertation (Master's Degree in Environmental Sciences), Postgraduate Programme in Environmental Sciences, Federal University of Tocantins, Palmas, 2011.

DWORKIN, R. **The Empire of Law**. São Paulo: Martins Fontes, 2003.

ELLIOTT, L. **The Global Politics of the Environment,** Houndsmills, Palgrave, 2002.

FANK, J. F. Notes on a complex legal-environmental epistemology for the information age In **Proceedings of the 2nd International Congress on Law and Contemporaneity**: media and rights in the network society, 2013, available at http://www.ufsm.br/congressodireito/anais accessed on 15/09/2016.

FARHI NETO, L. Concepções Filosóficas Ambientalistas In **ethic@,** Florianópolis, v.5, n.3, p. 33-56, Jul 2006.

FIGHERA, D. R. **The effectiveness of the ecological city project of Palmas (TO) through its green spaces.** Dissertation (Master's Degree in Environmental Sciences) - Postgraduate Programme in Environmental Sciences, Federal University of Tocantins, Palmas, 2005.

FIORILLO, C. A. P. **Curso de direito ambiental brasileiro**. 9ª ed. São Paulo: Saraiva, 2008.

FISCHER, R. M. B. **Foucault e a análise do discurso em educação** In Cadernos de Pesquisa, n° 114, p.197-223, November 2011, available at http://www.scielo.br/pdf/cp/n114/a09n114.pdf accessed on 05 August 2015.

FOUCAULT, M. **The Order of Discourse**. São Paulo: Edições Loyola, 1996.

____________. **Microphysics of power.** Organised and translated by Roberto Machado. Rio de Janeiro: Edições Graal, 2003.

Other Spaces". In: MOTTA, M. B. da (org.): Michel Foucault: Aesthetics: Literature and painting, music and cinema. São Paulo: Ed. Forense Universitária. 2001.

FREITAS, W. R. S.; JABBOUR, C. J. C. The case study(s) as a qualitative research strategy: foundations, application script and assumptions of excellence. In **XXX Encontro Nacional de Engenharia de Produção**, São Carlos, 2010.

GARCIA, D. S. S. Evolução Legislativa do Direito Ambiental no Brasil In **Boletim Jurídico**, Issue

752, December 2010, available at http://www.boletimjuridico.com.br/doutrina/texto.asp?id=2160, accessed on 24/03/2017.

GARCIAS, C. M. e BERNARDI, J. L. As funções sociais da cidade In **Revista de Direitos Urbanos**, Vol. 4 Curitiba: UNIBRASIL: 2008.

GENZ, K. S. **The Master Plan as an Urban Policy Instrument** available at https://www.mprs.mp.br/urbanistico/doutrina/id492.htm year2007, accessed on 14/08/2016.

GINZBURG, C. Representation: the word, the idea, the thing. In: **Wooden Eyes**: nine reflections on distance. São Paulo: Cia. das Letras, 2001.

HART, H. L. A. **O Conceito de Direito.** 2ª ed. Lisboa: Fundação Calouste Gulbenkian, 1996.

GOMES, P. C. C. **A condição urbana**: ensaios de geografia da cidade. Rio de Janeiro: Bertrand Brasil, 2002.

GONÇALVES, C. W. P. **Os descaminhos do Meio Ambiente**, São Paulo, Ed. Contexto, 2006.

GONÇALVES, R. C. **Responsabilidade Civil**. 8th ed., revised according to the new Civil Code. São Paulo: Saraiva, 2003.

GRINOVER, A. P. A **tutela dos interesses difusos**. São Paulo: Editora Max Limonad, 1984.

HACON, V. S.; LOUREIRO, C. F. B. Power structures and the environmental question: the reproduction of class inequality in **Revista DEMA,** vol. 29, April 2014 available at **http://revistas.ufpr.br/made/article/viewFile/33142/22437** access on 12/10/2016

HALUM, C. **Environmental Laws of Tocantins**. Palmas, Provision, 2009.

HARRÉ, R. J. **A Study of Environmental Discourse,** London, Sage, 1999.

HERTZBERGER, H. **Lições de Arquitetura**. São Paulo: Editora Martins Fontes, 1999.

HIDELBRAND, H. R. **The system as a work of art: complexity and ecosystems.** ANPAP, Belém, 2013 available at http://www.anpap.org.br/anais/2013/ANAIS/simposios/02/Hermes%20Renato%20Hild ebrand.pdf accessed on 15/01/2017.

LAGO, A. A. C. **Study 2 - International Negotiations on Climate Change Mitigation. Centre for Management and Strategic Studies in Science, Technology and Innovation**, Brazil. Thesaurus Editora, 2005.

_____________ . **Stockholm, Rio, Johannesburg: Brazil and the three United Nations environmental conferences.** Brazil. Thesaurus Editora. 2007.

LARAIA, R. B. **CULTURA** Um conceito antropológico 14.ed. - Rio de Janeiro: Jorge Zahar Ed., 2001.

LEFEBVRE, H. **O pensamento marxista e a cidade**. Lisbon: Ulisseia Publishing House, 1972.

_____________ . **The urban revolution**. Belo Horizonte: Editora da UFMG, 1999.

_____________ . **Space and politics**. Belo Horizonte: Editora UFMG, 2008.

LEFF, E. **Epistemologia Ambiental**. São Paulo: Cortez: 2002.

_____________ . Thinking about Environmental Complexity. In **Environmental complexity**. Blumenau, Edifurb, 2003.

LE PRESTRE, P. **International Ecopolitics.** São Paulo: Senac, 2000.

. **Environmental protection and international relations**: the **challenges of global politics, Armand Colin, 2nd edition, 2005.**

LIRA, E. R. **The genesis of Palmas-Tocantins**: the geopolitics of territorial (re)occupation in the Legal Amazon. Goiânia: Kelps, 2011.

LYRA, R. F. **What is Law?** São Paulo: Ed. Brasiliense: 2000.

LUHMANN, N. **La sociedad de la sociedad**. Mexico City: Herder. 2007.

KAMPHORST, M. A. John Rawls' Theory of Justice as Equity: a refutation of utilitarianism In **MICIMED** year 2014, available at https://www.imed.edu.br/Uploads/micimed2014 submission 175.pdf accessed on 15/01/2017.

KELSEN, H. **Pure Theory of Law.** São Paulo: Ed. Martins Fontes: 2009.

KLOEPFER, M. Towards the Environmental State. In **Umweltstaat**, Berlin, Heidenberg, 1989.

MACHADO, A. A. **International environmental: The social construction of the Bhopale extended chemical accident of ILO Convention 174**. Rio de Janeiro, vol. 28, no. 1, January/June 2006.

MACHADO, T. R. **The power of backwardness in the production of urban space:** the case of Palmas, Tocantins. Thesis (PhD in Geography), Department of Geography, Fluminense Federal University, Niterói. 2011.

MARQUES, P. R. V. **Industrial restructuring and the environmental issue**: a case study in a medium-sized company. Dissertation (Master's Degree in Regional Development and the Environment), Postgraduate Course in Regional Development and the Environment. Araquara University Centre, 2005.

MARUM, J. A. O. Meio Ambiente e direitos humanos In **Revista de Direito Ambiental,** São Paulo, v. 7, n.28, p. 116-137, Oct./Dec. 2002.

MARX, K. **Capital.** Critique of Political Economy, São Paulo: Editora Nova Cultural: 1996.

MEIRELLES, H. L**. Direito Municipal Brasileiro**. 6. ed. São Paulo: Malheiros, 1993.

MINAYO, M. C. S. **Pesquisa Social**: teoria, método e criatividade. Petrópolis: Vozes, 1994

MOL, A. **Globalisation and Environmental reform**: The Ecological Modernisation of the Global Economy, Cambridge, MA, MIT Press, 2003.

MORIN, E. **Introduction to complex thinking.** Porto Alegre: Ed. Sulina: 2007.

. **Land - Homeland**. Porto Alegre, Ed. Sulina: 2003.

_____________ . **Pour Entrer dans le siècle XXI**. France: Le Seuil, 2005.

MORGA. A. **ISO 14000 environmental management**. Pelotas, Faculty of Agricultural Engineering, UFPEL, 1996.

MOSCOVI, S. **The social representation of psychoanalysis**. Rio de Janeiro: Zahar, 1978.

MOURA, G. M. G. **Environmental Issues and Management Structure in Palmas (TO).** Dissertation (Professional Master's Degree in Regional Development) - Postgraduate Programme in Regional Development, Alves Faria University, Goiânia, 2011.

MUMFORD, L. **The City in History**: its origins, transformations and perspectives. São Paulo: Livraria Martins Fontes, 1998.

ODUN, E. P. **Fundamentals of Ecology.** São Paulo, Calouste Gulbekin Foundation, 2006.

OLIVEIRA, D. B; SILVA, A. S.;e et. al. A representação social do Meio Ambiente na leitura comunitária do Plano Diretor de Palmas (TO) In **Revista de Desenvolvimento e Meio Ambiente,** v. 35, p. 429-445, SER UFPR, dec. 2015.

UN. **Stockholm Declaration of 1972**. Available at www.mma.gov.br/estruturas/agenda21/files/stockholm.doc accessed on 12/09/2016.

_____________ . **Johannesburg Declaration on Sustainable Development** available at http://www.cqgp.sp.gov.br/gt licitacoes/publicacoes/joanesburgo.pdf accessed on 12/09/2016.

_____________ . **United Nations Conference on Development Sustainable - Rio+20**. Available at http://planetasustentavel.abril.com.br/blog/riomais20/2012/07/03/rio20-teve-grandes- results accessed on 12/09/2016.

PACHECO, R. S. **Actors and conflicts in urban environmental issues**. Espaço & Debates 35: 46-51. São Paulo, NERU, 1992.

PADILHA, N. S. **Constitutional Foundations of Brazilian Environmental Law.** Ed. Elsevier, 2010.

PALMAS **Organic Law of the Municipality of Palmas Law No. 00/90. Palmas** City Council, Palmas - Tocantins, 1990.

. **Organic Law of the Municipality of Palmas.** Law no. 48 of April 1990. Palmas City Council, Palmas, 1990.

. **Multiannual Plan**. Law No. 502 of 16 November 1994. Palmas City Council, Palmas, 1994.

. **Multiannual Plan**. Law No. 681 of 10 November 1997. Palmas City Council, Palmas, 1997.

. **Habite-se regularisation**. Law No. 721 of 31 March 1998. Palmas City Council, Palmas, 1998.

. **Municipal Environmental Policy**. Law No. 1011, of 04 June 2001. Palmas City Council, Palmas, 2001.

. **Friends of the Environment Programme**. Law No. 1023, of 05 July 2001. Palmas City Council,

Palmas, 2001.

. **Selective Waste Collection**. Law No. 1165 of December 2002.
Palmas City Council, Palmas, 2002.

. **Teaching the Environment**. Law No. 1189 of 27 May 2003.
Palmas City Council, Palmas, 2003.

______________ . **Palmas Urban Development and Housing Council.**
Law No. 1384, of 06 September 2005. Palmas City Council, Palmas, 2005.

. **Creation of Cesamar Park**. Law No. 1406 of 16 December 2005. Palmas City Council, Palmas, 2005.

. **Palmas Participatory Masterplan**. Complementary Law No. 155 of 28 December 2007. Palmas City Council, Palmas, 2007.

. **Permanent Education Forum**. Law No. 1911, of 10 August 2012. Palmas City Council, Palmas, 2012.

______________ . **I Municipal Education Conference on the Palmas Municipal Education Plan**: A Participatory, Democratic and Dialogical Construction.
November 2012, available at
http://www.palmas.to.gov.br/media/doc/arquivoservico/DOCUMENTO REFERENCI A PLANO MUNICIPAL DE EDUCACAO.pdf accessed on 10/11/2016.

. **Multiannual Plan**. Law No. 1946 of 31 December 2012.
Palmas City Council, Palmas, 2012.

. **Environmental Education Competition Award**. Law No. 1913, of 25 September 2012. Palmas City Council, Palmas, 2012.

. **Municipal Environment Foundation**. Law No. 2102, of 31 December 2014. Palmas City Council, Palmas, 2014.

______________ . **Single Annex to Law 2238** of 19 January 2016. Chamber
Municipal de Palmas, Palmas, 2016.

______________ . **SEMED minutes on the 2012 Municipal Education Plan**, available at http://www.palmas.to.gov.br/media/doc/arquivoservico/ accessed on 10/06/2016.

______________ . Finalisation of the Community Reading stage of the Plan revision
Director, available at **http://www.palmas.to.gov.br/secretaria/meio-ambiente/noticia/1503540/comissao-de-revisao-do-plano-diretor-de-palmas- finalizaa-etapa-leitura-comunitaria/** accessed on 10/11/2016.

______________ . Work Plan for the Revision of the Participatory Master Plan of Palmas
available at http://planodiretor.palmas.to.gov.br/ accessed on 10/11/2016.

______________ . Sustainable Palmas Programme. Palmas, Polis Institute, 2015.

PELUSO, M. L. O Desafio De Compreender A Natureza Na Obra De Milton Santos In **Revista Eletrônica: Tempo - Técnica - Território,** V.4, N.1, 2013, p. 21- available at http://inseer.ibict.br/ciga/index.php/ciga/article/viewFile/168/127 accessed on 15/11/2016.

PENA-VEGA, A. **The ecological awakening**: Edgar Morin and complex ecology. Rio de Janeiro: Garamond, 2003.

PESAVENTO, S. J. **História e História Cultural**. Belo Horizonte: Autêntica, 2003.

PIEPER, D. S.; SANTOS, T. F.; et. al. The collective, the social and the socio-environmental in representations through the discourse of the collective subject In **DELOS: Desarrollo Local Sostenible,** vol 5, n 15, October 2012, available at http://www.eumed.net/rev/delos/acercade.htm accessed on 15/09/2016.

PIVA, R. **Bem Ambiental.** São Paulo: Ed. Max Limonad: 2000.

PORTO-GONÇALVES, C. W. **Os descaminhos do Meio Ambiente.** São Paulo: Ed. Contexto, 1990.

PRADO, D. P. **Workers and the Environment: a study of industrial workers and their perception of the environment.** Rio Grande, Ed. FURG, 2001.

RANGEL, T. L. V. Principles of Environmental Law: the progressive construction of a diffuse right In **Âmbito Jurídico:** Rio Grande: XV, n. 99, April 2012: available at http://www.ambito-juridico.com.br/site/?n link=revista artigos leitura&artigo id=11357 accessed on 10/07/2015.

REALE, M. **Teoria do Direito e do Estado.** São Paulo: Ed. Saraiva: 2012.

REIGOTA, M. **O que é Educação Ambiental**. São Paulo; Brasiliense, 1994.

_____________. **Environment and social representation**. São Paulo, Ed. Cortez, 2007.

RICHARDSON et al. **Pesquisa social: métodos e técnicas**. São Paulo: Atlas, 1999.

ROBLES, G. **Law as text**: four studies in the communicational theory of law. Trad. Roberto Barbosa Alves. Barueri, Manole, 2005.

ROCHA, O. O **prefeito da transição,** relembra **Palmas** do **fim da década de 1990** available at http://www.t1noticias.com.br/cidades/odir-rocha-o-prefeito-da-transicao- relembra-a-palmas-do-fim-da-decada-de-90/58037/ accessed on 20/10/2016.

SACHS, I. **Caminhos para o desenvolvimento sustentável**. Rio de Janeiro: Garamond, 2002.

. **Transition Strategies for the 21st Century -** Development and the Environment. São Paulo: Studio Nobel, 1993.

SANTOS, B. S. **The Critique of Indolent Reason**. São Paulo, Cortez, 2011.

SANTOS, M. **Espaço e método**. São Paulo: Studio Nobel, 1985.

_____________. **Technique, space and time.** Globalisation and the scientific-informational technical environment. São Paulo, Hucitec, 1994.

_____________. The rediscovery of Nature. In **Estudos Avançados**, 6(14), 1992, p. 1-15 available at http://www.scielo.br/pdf/ea/v6n14/v6n14a07.pdf accessed on 23/11/2015.

_____________. Ética ambiental e função do direito ambiental In **Revista de Direito Ambiental**, São Paulo, v. 18, Apr.-Jun. 2000.

_____________. **The nature of space**, Edusp, São Paulo, 2002.

SARLET, I. W. **The Effectiveness of Fundamental Rights**. Porto Alegre: Livraria do Advogado, 1998.

SILVA, A.; MACHADO, G., COELHO, N.; MARLA, P. **Environmental Education at the Santa Barbara Full-Time Municipal School, in the municipality of Palmas-TO.**
Palmas, Catolica TO, 2011, available at http://www.catolica- to.edu.br/portal/portal/downloads/docs gestaoambiental/projetos2011-1/1- período/EDUCACAO AMBIENTAL NA ESCOLA MUNICIPAL DE TEMPO IN TEGRAL SANTA BARBARA NO MUNICIPIO DE PALMAS-TO.pdf accessed on 05 /08/2015.

SILVA, J. A. **Constitutional Environmental Law**. 7ª edition. São Paulo: ed. Malheiros, 2009.

SILVA, L. H. P. O ambiente como campo e a justiça ambiental à luz da teoria do poder simbólico de Pierre Bourdieu In **Cabo dos Trabalhos**, n. 7, 2012 Electronic journal of doctoral studies at the Centre for Social Studies, University of Coimbra available at **https://cabodostrabalhos.ces.uc.pt/n7/documentos/02-LaysHelenaPaeseSilva.pdf** accessed on 12/11/2016.

SILVA, O. B. **Breve Historia do Tocantins e Sua Gente -** uma luta secular.
Araguaina, solo editores, 2006

SILVA, O. F. **Environmental Law and Ecology**: contemporary philosophical aspects. Buaeri, Manoli. 2003.

SILVA , V. C. P. **Palmas, a Última Capital Projetada do Século XX**: uma cidade em busca do tempo, São Paulo, Ed. UNESP, 2010.

SPRINGER, K. S. A Concepção De Natureza Na Geografia In **Mercator**, vol 9, n 18, 2010, available at http://www.mercator.ufc.br/index.php/mercator/article/viewFile/368/270 accessed on 15/11/2016.

SOUSA, A. L. O Território da Habitação no Processo do Plano Diretor Participativo da Cidade de Fortaleza/Ce, Brasil. In **Scripta Nova**. Revista Electrónica de Geografia y Ciencias Sociales. Barcelona: Universidad de Barcelona, 1 August 2010, vol. XIV, n° 331 (9). Available at http://www.ub.es/geocrit/sn/sn-331/sn-331-9.htm accessed on 02 /12/2016.

SOUZA, H. et al. **Narrow knots: memories of a sower of utopias.** Brazil. Editora Garamond. 2001.

SOUZA, M. L. **A prisão e a ágora**: Reflexões em torno da democratização do planejamento e da gestão das cidades. Rio de Janeiro: Ed. Bertrand Brasil: 2006.

SOUSA, R. M. - A expansão urbana do plano diretor de palmas: uma análise jurídica acerca da disputa **do** espaço urbano em Palmas/TO In **Vertentes do Direito**, ano 01, edição 03, 2015, available at https://www.google.com.br/url?sa=t&rct=i&q=&esrc=s&source=web&cd=1&cad=rja&uact=8&ved=0ahUKEwiEk9329OTRAhUCxpAKHYO9A1EQFggcMAA&url=https%

3A%2F%2Fsistemas2.uft.edu.br%3A8004%2Findex.php%2Fdireito%2Farticle%2Fview%2F1526%2F8372&uss=AFQiCNESWKgsc7xzuGtiXqUvSX9BBu8QQA&sig2=3P ZjG4jRBse 0gn m8VpjQ accessed on 15/10/2016.

SUITER, H. N. **O Meio Ambiente e o pensamento sistêmico complexo,** 2009, available at http://www.webartigos.com/artigos/o-meio-ambiente-e-o-pensamento- sistemico-complexo/21237 accessed on 07/09/2016.

TARDIF. M. **Saberes Docentes e formação profissional,** Petrópolis, editora vozes, 2002.

THEMAG ENGENHARIA. Environmental Impact Study, volume IV - environmental programmes and conclusions - November 1996 available at http://www.edp.com.br/geracao-renovaveis/geracao/tocantins/investco/empresa/documentos-oficiais/eia- rima/Documents/8vol IV programas ambientais conclusoes.pdf accessed on 20/10/2016.

TOCANTINS. **Environmental Policy of the State of Tocantins.** Law No. 261/91. Legislative Assembly of the State of Tocantins: Tocantins, 1991.

. **State** Water **Resources Policy**, Law No. 1.307/02. Legislative Assembly of the State of Tocantins; Tocantins, 2002.

. **State Environmental Education Policy**. Law No. 1374/03. Legislative Assembly of the State of Tocantins; Tocantins, 2003.

___________ . **State Education Plan**. Law 1859 of 06 December 2007. Legislative Assembly of the State of Tocantins: Tocantins, 2007.

. **Tocantins State Environment Council.** Law No. 1.789/07. Legislative Assembly of the State of Tocantins: Tocantins, 2007.

___________ . **Climate Change, Environmental Conservation and Sustainable Development in Tocantins**. Law No. 1.917/08. Legislative Assembly of the State of Tocantins: Tocantins, 2008.

. COEMA/TO **Minutes No. 06/1996 of 28/02/1996** available at https://central3.to.gov.br/arquivo/246832/ accessed on 14/12/2016.

TEIXEIRA, L. F. C. **The formation of Palmas.** Revista UFG, Ano XI, n 6, Goiânia, 2009.

TOZONI-REIS, M. F. C. Training environmental educators and paradigms in transition In **Ciência & Educação**, v.8, n°1, p.83 - 96, 2002.

VELASQUES, A. B.; MACHADO, T. R. **O Pensamento Urbanístico na Concepção de Novas Cidades no Brasil:** permanências e rupturas no projeto de Palmas, Tocantins. VXI ENAPUR, Belo horizonte, 2015 available at http://docplayer.com.br/26276093- O-pensamento-urbanistico-na-concepcao-de-nas-cidades-no-brasil-permanencias-e- rupturas-no-projeto-de-palmas-tocantins.html accessed on 14/09/2016.

VELASQUES, A. B. A Concepção de Palmas(1989) e sua Generalidade: Urbanismo e Capitalidade na História **da** Criação de novas cidades In **Xiv Encontro Nacional Da ANPUR**, 2011 Rio de Janeiro available at http://unuhospedagem.com.br/revista/rbeur/index.php/anais/article/viewFile/3346/3279 accessed on 15/11/2016.

VELASQUES, A. B.; RANGEL, G. B. Memória Urbana de Palmas-TO: Levantamento de Informações e Material sobre o Plano de Palmas e seus Antecedentes In **Seminário de Iniciação Científica UFT**, Palmas, 2012 available at http://eventos.uft.edu.br/files/imports/viii cient/documentos/1ea21cd52b2251b1637783 89c2aa0b09/1689.pdf accessed on 15/11/2016.

UNESCO. **Earth Charter.** United Nations Organisation 2002, Ratified by UNESCO. Available at http://rebal21.ning.com/page/a-carta-da-terra accessed on 12/09/2016.

___________ . **Education for a sustainable future**: a transdisciplinary vision for shared actions. Brasília: Ed.IBAMA, 1999.

UNGER, N. M. **O Encantamento do Humano - Ecologia e Espiritualidade.** 2ª ed., São Paulo, Edições Loyola, 2000.

XAVIER, F. O. **Palmas: A Capital for All?** Dissertation (Master's in Geography) - Postgraduate Course in Geography, Federal University of Paraná, Curitiba, 2007.

Printed by Books on Demand GmbH, Norderstedt / Germany